HUBRIS

ALSO BY ALISTAIR HORNE

But What Do You Actually Do?

Kissinger: 1973, The Crucial Year

The French Revolution

The Age of Napoleon

La Belle France

Seven Ages of Paris

Telling Lives (editor)

How Far from Austerlitz? Napoleon 1805–1815

The Lonely Leader: Monty, 1944–45 (with David Montgomery)

A Bundle from Britain

Harold Macmillan: Volume I, 1894–1956

Harold Macmillan: Volume II, 1957–1986

The French Army and Politics, 1870–1970

Napoleon, Master of Europe, 1805–1807

A Savage War of Peace: Algeria, 1954–1962

Small Earthquake in Chile

The Terrible Year: The Paris Commune, 1871

Death of a Generation

To Lose a Battle: France, 1940

The Fall of Paris: The Siege and the Commune, 1870–71

The Price of Glory: Verdun, 1916

Canada and the Canadians

The Land Is Bright

Back into Power

18 17
HARPER

An Imprint of HarperCollins*Publishers*

HUBRIS

THE TRAGEDY OF WAR IN THE TWENTIETH CENTURY

ALISTAIR HORNE

HarperCollins books may be purchased for educational, business, or sales promotional use. For information, please e-mail the Special Markets Department at SPsales@harpercollins.com.

Originally published in Great Britain in 2015 in a different form by Weidenfeld & Nicolson.

Portions of this book originally appeared in a different form in *Military History Quarterly.*

FIRST EDITION

Designed by William Ruoto
Maps by John Gilkes
Frontispiece: Korea Civil War, © Hank Walker/Getty Images

Library of Congress Cataloging-in-Publication Data has been applied for.

ISBN: 978-0-06-239780-5

15 16 17 18 19 OV/RRD 10 9 8 7 6 5 4 3 2 1

For Nicholas Berry, acute student of military history,
and the most supportive of patrons, in the finest tradition

I'm not a theoretical historian, seeking to guide the reader to a general conclusion. I'm quite content to be a narrative chronicler, a slave of the facts.

—Sir Martin Gilbert, 1996

CONTENTS

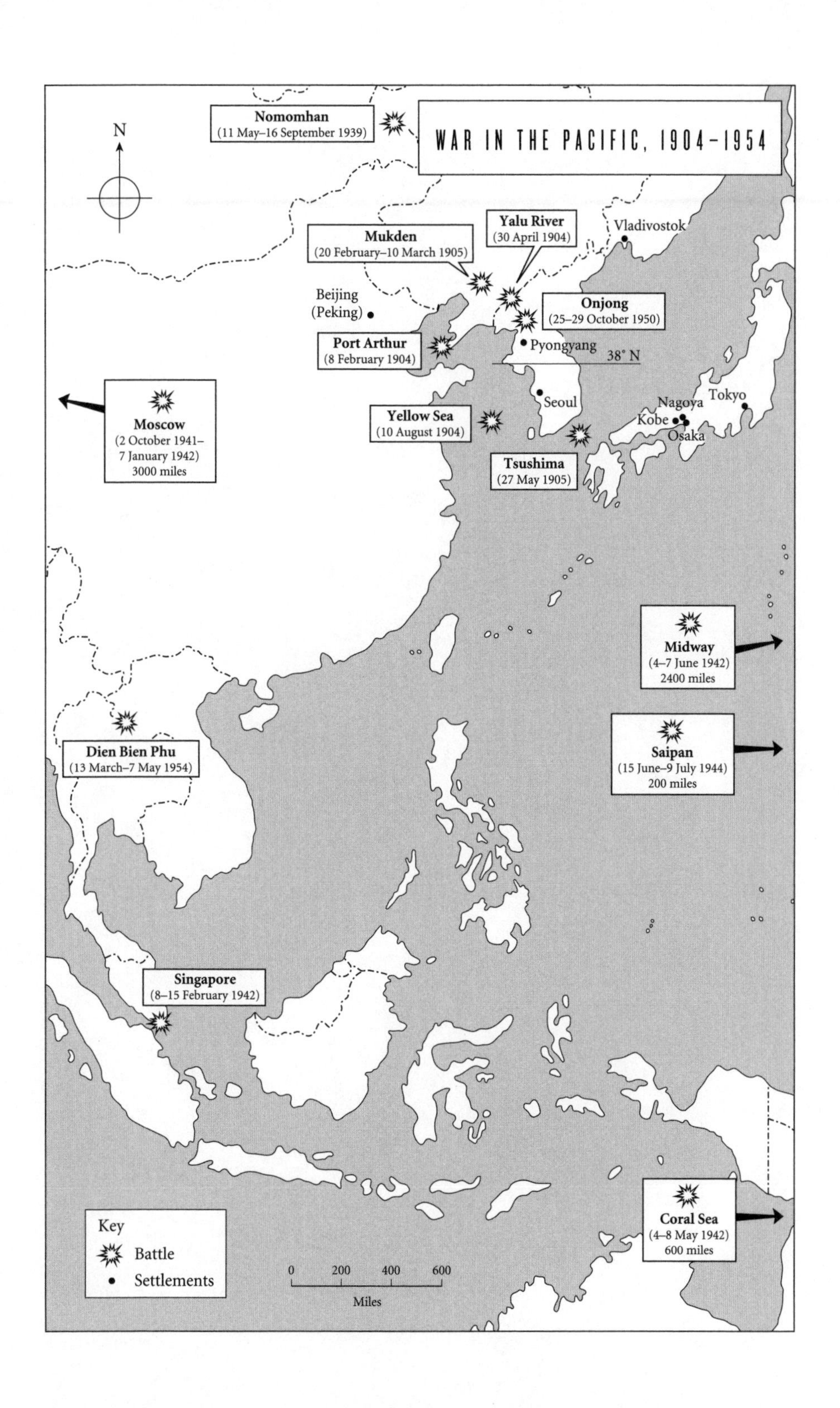

WAR IN THE PACIFIC, 1904–1954
N
Nomomhan
(11 May–16 September 1939)
Mukden
(20 February–10 March 1905)
Yalu River
(30 April 1904)
Vladivostok
Beijing
(Peking)
Onjong
(25–29 October 1950)
Port Arthur
(8 February 1904)
Pyongyang
38° N
Seoul
Nagoya
Tokyo
Kobe
Osaka
Moscow
(2 October 1941–
7 January 1942)
3000 miles
Yellow Sea
(10 August 1904)
Tsushima
(27 May 1905)
Midway
(4–7 June 1942)
2400 miles
Dien Bien Phu
(13 March–7 May 1954)
Saipan
(15 June–9 July 1944)
200 miles
Singapore
(8–15 February 1942)
Coral Sea
(4–8 May 1942)
600 miles
Key
Battle
Settlements
0
200
400
600
Miles

Prologue

THE ANCIENT GREEKS defined hubris as the worst sin a leader, or a nation, could commit. It was the attitude of supreme arrogance, in which mortals in their folly would set themselves up against the gods. Its consequences were invariably severe. The Greeks also had a word for what usually followed hubris. That was called *peripeteia*, meaning a dramatic reversal of fortune. In practice, it signified a falling from the grace of a great height to unimaginable depths. Disaster would often embrace not only the offender, but also his nearest and dearest, and all those responsible to him.

Having written, over the course of fifty-odd years, numerous books and articles on warfare in its various shapes, I sat down some time ago to reflect on what might be its common features that stand out over the ages. One that emerged preeminently was hubris: wars have generally been won or lost through excessive hubris on one side or the other. In modern military parlance it might also be dubbed "overreach." So this is the genesis of the current work. Wars and battles seldom happen in isolation, in a vacuum. Each has its causes from the past, and each has its often baneful consequences in a subsequent period. To study them, the good historian needs to be able to scan backward and forward, as well as sideways. Thus I have focused on those conflicts that affected future history powerfully in ways that transcended the actual war in which the conflict was set.

I chose to limit my study to the first half of the twentieth century, the bloodiest century in history, and a century that indeed could be called the century of hubris—during which humans were slaughtered in numbers to exceed those of any other century, all at the whim of one or two warlords or dictators.

One immediate effect of hubris is often complacency, a first step on the path to ruin. As the German chancellor Otto von Bismarck once remarked, in an axiom that might be seen as a prediction of the eventual fate of his own country, "A generation that deals out a thrashing is usually followed by one which receives it."

Two battles that I have already written about, the German Siege of Paris (1870–1) and the gory Battle of Verdun in World War I (1916), provide good examples of the validity of Bismarck's warning. Out of their victory in 1871, the Germans emerged so well stuffed with arrogance that defeat in the next war, if not the one after, was all but certain. Nearly a half century later, the French heirs to the terrible pyrrhic victory of Verdun on the one hand felt that they could never repeat such a sacrifice and, on the other, were impregnated with the hubristic self-confidence that they would be safe behind the super-Verdun-like fortress of the Maginot Line, and that the hereditary German foe *ne passera pas* (shall not pass). The shibboleth of their fathers' heroism was enough. Or was it? Six weeks in the summer of 1940 would prove that it was not.

This book is divided into five parts. I begin with the 1905 Battle of Tsushima, and the Japanese sinking of the Russian fleet at Port Arthur—an event that shocked the world. Bracketed within that is a look at the triumph of Japanese ambitions. In the second part, the little-known Battle of Nomonhan in Mongolia in 1939 illustrates the rise of Soviet power and the first check on Japan. Third, after Hitler had overreached himself with his invasion of the Soviet Union, there is the Battle of Moscow in 1941, which takes us to Pearl Harbor and to the fourth part, the turning point of the Pacific War with the Battle of Midway. Key episodes from the Korean War, in part 5, provide a perfect illustration of the hubristic folly of not knowing when to stop. I end with the French disaster in Vietnam at Dien Bien Phu, the last of the old-fashioned colonial wars in the Far East.

My choice of subjects may well be challenged as capricious: it is certainly idiosyncratic and personal. Deliberately, the First World War is left out. It seemed to me that the whole war began, and was caused by, various sublime practitioners of hubris in conflict with one another. Further, it would be difficult to identify any one battle that held calamitous consequences for the future. The whole war did that.

I would hesitate to write anything to belittle British prowess in either world war. But where was the battle in which hubris displayed by the leadership affected postwar events? El Alamein? Caen? Arnhem? Field Marshal Bernard Law Montgomery was certainly a candidate for hubris in the eyes of his allies. But, as I tried to show in my biography *The Lonely Leader*, Montgomery had his special reasons. He inherited a battered army that had been defeated almost incessantly since the beginning of the war and was justifiably alarmed by the Wehrmacht; in consequence, he had to infuse it with large doses of what he called "Binge"—the right spirit for victory. Then again, without belittling British arms, the Battle of El Alamein, designated by Churchill as "the end of the beginning," was indeed a small affair compared numerically with the troops arrayed before Moscow in 1941—over ten to one in comparison with El Alamein—and could there be found in any of Monty's battles issues that would affect postwar history?

To my mind, the Battle of Moscow was more definitively an "end of the beginning," and probably more than any other salient victory, it was to have an influence on Soviet conduct in postwar events. Even today, the scale of the fighting and the numbers before Moscow in 1941 stun the imagination. Seen from this distance, it was a true turning point in the war, more significant even than Soviet marshal Georgy Zhukov's great victory the following year at Stalingrad. It marked a decisive moment in warfare, as the first time that the apparently invincible Panzers were stopped, defeated, and then forced to retreat. It also marked Hitler's final loss of belief in his generals, displaced by his belief in his own star—with the disastrous consequences that remain familiar to history. More than that, the victory, and its cost to the Russian people, established in Stalin's mind the shape of the postwar map of Europe under a Soviet aegis. More immediately, it confirmed him as the irreplaceable, omnipotent Russian war leader. As far as overall Allied strategies were concerned, so much after Moscow seems to have been dancing to Stalin's tune. Certainly, for the next unhappy forty-five years, the shape of Europe would be the shape that Stalin had dictated.

Much of this book concerns the Pacific and Japan. It's an area I've not much written about before. Perhaps I may plead, in part, the enticement for a historian, and the sheer excitement, of uncovering fresh green fields, their consequence to world history perhaps not adequately explored. Over the years, I have written, in one form or other, about most of the battles in this book, from Verdun and Dien Bien Phu to Pearl Harbor and Korea. But I have never before studied the Japanese side of things in depth. It has been instructive. My argument is that this is where it all began, with Admiral Togo's far-off victory over the Russians at Tsushima in 1905. In their invasion of Manchuria in 1931, Japan's warlords showed the way to aggression two years before Hitler came to power. A great deal subsequently flowed from this, in terms of the huge imbalance it caused in world affairs. And, of course, it was in the Pacific, within a few miles of Tsushima, that the sword of Damocles descended on Japan with such catastrophic force in 1945. If one can read the tea leaves correctly, the Pacific theater may well be the arena for future disputes between the major world players.

Characters in history often carry within them hooks and eyes that can provide a certain linkage. For instance, the Japanese commander Isoroku Yamamoto, who as a lieutenant after the Battle of Tsushima went to comfort the humiliated Russian admiral in his hospital bed, would be the leader appointed to inflict a copycat defeat on the American fleet a generation later at Pearl Harbor. At the Battle of Midway in 1942, Yamamoto would be roundly chastised for his hubris in attacking Pearl Harbor six months previously. Midway clearly marked a peripeteia in Japan's imperial pretensions, a harbinger of the doom that would overtake the country three years later. With parallel congruence, one of the US leaders most responsible for Yamamoto's eventual defeat, General Douglas MacArthur, would live to see a hitherto triumphant career plunged into disgrace following one act of hubris in the ensuing Korean War.

As one who would mete out the punishment prescribed by hubris, Zhukov, on the Mongolian battlefield of Nomonhan in 1939, would emerge as one of the Soviet leaders responsible for beating back Hitler's armies on the outskirts of Moscow in December 1941. Zhukov, appropriately, would then go on, through the triumph of Stalingrad, to inflict the ultimate destruction of the Führer's evil dreams in the ruins of Berlin.

Thus, all the conflicts I have chosen have links with one another, as do many of the players. One of the most depressing facts about military history is how very little the great warlords ever learn from the mistakes—indeed, the calamities—of their predecessors. A thread running all the way through my selection is a kind of racist distortion whereby one power persists in writing off its foes because of the color of their skin or the slant of their eyes, or the supposed backwardness of their culture. Thus we may note the Russians' contempt for the little yellow men; and, in reverse, the Japanese tendency to relegate the Chinese to the rank of *Untermenschen* (subhuman), much as Hitler regarded the barbarian Slavic hordes during Operation Barbarossa. Before and even after December 7, 1941, American indoctrination was persuading pilots that the "Japs" couldn't fly because of their poor eyesight. It was a racial legend that would pursue the United States into the Korean War, with the denigration of the "dumb gooks," and on into Vietnam. That sturdy notion of occidental superiority would influence the French defenders of Dien Bien Phu and even beyond: the heroic survivors of the final colonial debacle in Southeast Asia would carry the fatal mystique on into the Algerian War.

One of the incidental questions prompted by this study is: Why don't successful generals know when to stop? A good general should know this, unlike those of the western front constantly battering away at an illusory target regardless of the human cost. How different world history might have been if MacArthur had had the good sense to stop on the Thirty-Eighth Parallel. The answer of course is that it is hubris itself that blinds generals. But we students of history should not succumb to our own arrogance in supposing that hubris is easy to avoid. It arises out of success. In the aftermath of triumph, anything seems possible. And that, as this book tries to show, is when so many calamitous decisions are made. If a leader is successful, why hold him back? This book tries to provide an answer.

PART ONE

Tsushima, 1905

CHAPTER 1

The New Century

AS NEW CENTURIES are wont to do, the opening years of the twentieth seemed generally full of promise—of a continuation of Victorian peace, prosperity, and progress. Certainly there was no hint that might lead the most pessimistic of Cassandras to predict the horrors that would lie ahead, making it the most savage century in the history of mankind. Back in another untroubled summer, that of 1870, the British foreign secretary Lord Granville, gazing up from Whitehall, could detect "not a cloud in the sky." Yet a month later, Europe would be torn asunder by the Franco-Prussian War, marking the end of a century of Pax Britannica and all its optimistic assumptions. But this was just the sort of dirty trick that history plays to confound historians—and the architects of grand policy.

But the twentieth century had also opened on a note of melancholy and uncertainty. Britain, as the Victorian age ran to its end, was subjected to unheard-of humiliations at Ladysmith and Spion Kop in the Boer War, which had broken out the previous October. It would be a year before the British military could reassert its superiority. In the Far East, fear and confusion among Europeans spread with the outbreak of the Boxer Rebellion, while in the mid-Pacific the new American imperialism put down its marker by annexing Hawaii.

In England, on January 22, 1901, the old queen, who had stamped her name on most of the previous hundred years, died. The uncertainty in every Briton's mind was whether, without Victoria, things would go on as before, pursuing their same comfortable course. Reassuringly, a vast area of the inhabitable globe was tinted a friendly English pink; in terms of mass, the only area to compete, generally shaded a cold green, was the Russian colossus, which stretched unmanageably from the eastern frontiers of Germany to Vladivostok. But apart from the little-seen skirmishes in the "Great Game" regularly taking place on the barren fringes of Central Asia where the pink and the green met, backward Russia presented little threat. Of comfort was the fact that, almost as the old queen's life ebbed to its close, the unpleasant Boer War was ending. Britain, having suffered a series of shocking defeats inflicted on her regulars by a bunch of armed farmers, had emerged with no distinction, moral or military. Humiliating reverses had been dished out with an ease similar to that administered at Concord, Massachusetts, over a century earlier. Jealous nations among Kipling's "lesser breeds," such as Kaiser Wilhelm II's Germany, could not help sitting up and taking notice. And so too, across the world, on the fringe of China and the Pacific, did the newly emerged nation of Japan.

In June 1900, the European powers, comfortably established in their concessionary enclaves carved from the decaying body of China, were rocked by the outbreak of the Boxer Rebellion. A sudden eruption of the suppressed Chinese proletariat, outraged by the unequal treaties that the "foreign devils" had imposed and led by the so-called Society of Righteous and Harmonious Fists, massacred hundreds of Europeans in Peking (Beijing), including the German ambassador. Overstretched in Africa, Britain found itself having to fall back for help on a new ally, imperial Japan, an unknown quantity only recently released from its centuries-long hibernation behind its self-imposed lacquer screen. Within a few months, the Boxer Rebellion's leaders were executed in Peking and the rebellion ended with the signature in September 1901 of the Peking (or Boxer) Protocol, which permitted the "powers" to resume their bad old greedy ways. The difference was that now there was a new player on the scene: Japan. Some with the gift of prognostication might have deemed that the lid of Pandora's box had been lifted.

In terms of the technology of warfare, though undetected at the time, there were certainly pointers in the century following the epic struggle against Napoleon with which the nineteenth century had begun. There was the American Civil War, as well as various minor wars, to suggest what modern soldieries could do to each other on a land battlefield. But since Trafalgar in 1805, there had been no major battle at sea to suggest that warfare there too may have evolved. Ever since the invention of the cannon and its installation aboard ship—since before the Spanish Armada of 1588—the basics of naval warfare had remained little changed. Great wooden ships, studded with massive guns and propelled by acres of sail, hammered away at the enemy at almost point-blank range, from three hundred yards at most, until one or the other was reduced to a mastless hulk, or blew up. It was all about the weight of the broadside. Tactics too had little altered; every midshipman would dream of one day "crossing the T" of an enemy column, as Nelson had done at Trafalgar, maneuvering a line of ships to sail across the front of the Franco-Spanish fleet and so enabling the British to fire broadsides while the enemy could deploy only his forward guns.

Yet there was one engagement, a very minor and inconclusive one, that gave an indication that, although out of sight, a most fundamental change in naval warfare might be under way. That, too, took place in the Western Hemisphere, right at the beginning of the American Civil War. At Hampton Roads, Virginia, on March 9, 1862, two strange-looking craft, contoured like horseshoe crabs, one called the *Merrimack* and the other the *Monitor*, engaged each other. They were the first "ironclads" in the history of naval warfare. They hammered away at each other for about three hours without being able to inflict significant damage, then backed off. The two American ironclads did not fight again, but a notable point had been reached. The days of the "sides of oak" were henceforth numbered. Vanished almost overnight were the awe-inspiring ships of the line like Nelson's *Victory* or the lithe greyhound-like frigates that had harried Britain's Royal Navy during the War of 1812. Navies the world over hastened to refurbish their fleets at immense cost. The two major naval powers, Great Britain and France, had already halted construction of wooden-hulled ships, and others followed suit. In 1859 France launched a new iron supership, properly called *La Gloire*, the world's first oceangoing ironclad, a steam-driven behemoth of 5,630 tons and a crew of 570. The next year, Britain followed up with a mammoth twice its size, HMS *Warrior*.*

The world's first arms race was on. Fortunately there was no conflict in sight to lend it particular urgency, and neither *Warrior* nor its French rival would ever see action. Navies would from now on protect the vitals of their ships with great slabs of cast iron of ever-increasing thickness. Even more formidable was the development of the long-range, rifled, breech-loading guns of monster calibers that could hit and destroy an enemy ship with huge, high-explosive shells at seven thousand and even ten thousand yards' range, the guns themselves now protected in revolving armored turrets—no more the murderous eyeball-to-eyeball grappling of Nelson's day. In October 1901 the Royal Navy launched its first submarine, a weapon that was to prove deeply disadvantageous to the progenitor nation.

If it challenged the navies of the established world powers to compete, the naval race also offered newcomers an entrée hitherto closed to Kipling's lesser breeds. One such was Kaiser Wilhelm II's Germany, and another Meiji Japan.

Nevertheless, to the amazement of most of the world, not just the injured party, this peaceful scene was scarred on February 8, 1904, when Japan launched a surprise attack on the Russian base of Port

*Superbly reconditioned, the *Warrior* lives on today in Portsmouth Harbor.

Arthur (or Lüshun, as it was known by its previous, and perhaps rightful, owner, China), at the tip of a Manchurian promontory in the Yellow Sea. Three weeks later there followed reports of Russian troops retreating from Korea into Manchuria, pursued by an army of one hundred thousand Japanese.

Initially, not much occurred to upset the applecart in the rest of the world. Then, suddenly, on October 22, Great Britain was rocked by news that the Russian Baltic Fleet, appearing near the Dogger Bank off the northeastern coast of England, had attacked a British fishing fleet and sunk a trawler, killing members of its crew. The Russians' excuse was that they had mistaken the trawlers for Japanese torpedo boats that they believed were hunting them. Japanese torpedo boats in the North Sea? There was consternation and outrage in the clubs of Saint James's, where Britain's power elite congregated. Calls were made to declare war instantly on Russia, or at least to demand extensive reparations. But what *was* this war all about? Who were they, these Japanese, anyway? And what did they think they could possibly achieve against the might of imperial Russia, all that green on the map stretching from the Pacific to the Vistula? Where had these Japanese come from?

What most in the West knew of Japan was more or less limited to its skill in the decorative arts. Hokusai's woodblock print *The Great Wave*, for example, influenced the French impressionists from Manet to Renoir, and art nouveau artists in Germany. There were also netsuke, but awareness of these miniature sculptures was limited to the more refined occidental collectors. In 1885 there came Gilbert and Sullivan's *The Mikado; or, The Town of Titipu*, a rather cruel send-up of pre–Meiji Restoration Japan. Though it was intended as a farce on contemporary England, some Japanese critics saw the setting of a medieval Japan as deeply disrespectful of the revered, and very modern, Emperor Meiji. Then came Puccini's *Madama Butterfly*, another work hardly flattering of Japan's harsh social customs, and equally offensive to its modern sensibilities. (*Butterfly*'s first night at La Scala took place within days of the outbreak of the Russo-Japanese War; it was a major flop.)

If Western views of Japan at the turn of the century were not flavored by ignorance and condescension, they were downright contemptuous. Tsar Nicholas II dismissed the Japanese as "little yellow men from whom Europeans have nothing to fear," or simply as "monkeys." His cousin and fellow monarch Kaiser Wilhelm II, ever the provocateur, sympathized with his views on the "Yellow Peril": it was Russia's divine mission "to defend Europe from the inroads of the Great Yellow Race." The views privately expressed in the clubs of Saint James's would probably have been not dissimilar.

Nevertheless, here was a small nation, imbued with extraordinary discipline and an aggressive spirit, which had emerged only recently from self-imposed obscurity. Tucked away in the isolation of its island strongholds, the Japanese nation had, from time immemorial, developed its own myths and beliefs, alien and largely incomprehensible to the rest of the world. To begin with, whereas monarchs in medieval or Renaissance Europe could claim to be God's Anointed, the emperor of Japan was divine in his own right.

Thus, ordinances issued by ministers in the emperor's name had to be regarded as coming directly from God. This would go some way to explain the fanatical bravery of Japanese combatants through the ages. By legend, the first Japanese were descended from the brother of the Sun Goddess, a contentious and violent figure called Susanoo-no-Mikoto. In a direct line from him came the first emperor, Jimmu Tenno, who, after a great deal of combat and bloodshed, set up court in the central province of Yamato. The establishment of his throne, often given as occurring in 660 BC, is still celebrated to this day, and is thereby according to the current imperial dynasty by far the longest bloodline of any royal family in history.

Through the ages Japan tended to regard its neighbors with a mixture of fear and covetousness. The huge, shambling mass of China just across the East China Sea was always its principal worry, real or imagined. Twice in the thirteenth century Japan was threatened with invasion by the dread Mongol hordes of Kublai Khan, operating out of a vassal China. On the second occasion, in 1281, the Mongols had amassed an army estimated to be 150,000 men strong. They established a toehold on Kyushu Island. Then, suddenly, the Mongol fleet was hit by a typhoon and virtually destroyed. Japanese history recorded it as the "Divine Wind," or Kamikaze. It was a term that would gain fateful significance in the latter, desperate days of the Second World War, as the name accorded to the imperial suicide bombers who attacked the US fleet. But, historically, the typhoon reinforced the legend of the nation's divine origin, something that would always protect it when confronted by disaster.

To meet foreign threats like the Mongols, Japan built up its navy, a tradition that would survive through the centuries, and one similar to that of another island kingdom, Britain. In 1592, Japan used its navy to invade Korea, possibly as a prelude to attacking China proper. There followed a bitter campaign, lasting six years, and the establishment of an appallingly harsh and cruel regime that left an imprint of enduring hatred among the Koreans, who also would never forget the name of the daimyo or feudal warlord Hideyoshi.

Inevitably, with the advent of the Renaissance in Europe, the West came in contact with Japan. First were Portuguese merchants and Jesuit priests. Shortly after his arrival in Japan in the sixteenth century, Father Francis Xavier wrote effusively: "The people whom we have met so far are the best who have yet been discovered, and it seems to me that we shall never find among heathens another race to equal the Japanese. They are people of very good manners, good in general, and not malicious; they are men of honour to a marvel, and prize honour above all else in the world."

Perhaps the good father was lucky in his contacts. At any rate, the honeymoon lasted some fifty years, reaching a point where the Jesuits seemed to have good prospects of converting the whole of Japan to Christianity. Then, by the end of the sixteenth century, a reaction set in. Twenty-six Christians were martyred by the daimyo tyrant Hideyoshi—acting, of course, in the name of the emperor. English and Dutch traders who had set up commerce with Japan became almost as unpopular as these Christians among the Japanese. By the 1630s, Japan's rulers decided that they had had enough of foreigners and their interference in domestic affairs. Abruptly all interaction with the outside world was cut off. In 1647, a draconian edict decreed that any Japanese leaving the country would do so under pain of death, and the same fate would await him if he returned. Foreigners attempting to enter would be similarly at risk. So Japan's self-isolation continued for two centuries, with consequences felt until the mid-twentieth century. On the other hand, as the historian Richard Storry remarks, "the Japanese lived in peace . . . with themselves and with the world, for two and a half centuries—a record that most nations, reviewing their own history over a similar period, must surely envy."

At the same time, an unpleasant manifestation known as *Bushido* took root among the samurai. Achieving a semireligious status, Bushido was best described as "a Spartan devotion by a warrior class to the arts of war, a readiness for self-sacrifice." Its full unpleasantness would be demonstrated in the Second World War, and before that in the invasion of China.

The aim of the ruling Tokugawa shogunate was, simply, to maintain the status quo of antiquity. But then, in the mid-nineteenth century, came a rude awakening. On July 8, 1853, four sinister black warships, commanded by US commodore Matthew Perry (and accompanied by Puccini's fictional Pinkerton), forced their way into Uraga Harbor, near Edo (modern Tokyo). Two of Perry's squadron were steamships, a novelty never before seen in isolated Japan. The commodore rejected demands to leave, insisting instead that he present a letter from President Millard Fillmore, who probably did not know exactly where Japan was. Perry threatened force if the Japanese resisted. The presidential letter contained a demand that the Japanese regime open up trading relations—the first display of "Yankee Imperialism." Perry announced that he would return the following year for an answer. He was back in February, this time with seven warships, but in the interval the ruling shogunate had undergone a remarkable about-face. Antiquated Japan was more than ripe for change, and by 1857 Japan was telling the United States, "Intercourse shall be continued forever." Two centuries of isolation was abandoned almost overnight. A new, modern-minded oligarchy thrust aside the feuding daimyos and the Tokugawa shogunate, restoring centralized power to the new emperor, the fifteen-year-old Mutsuhito, in 1867. He took on the name of Meiji, meaning "enlightened rule." Historians of Japan would dub the astonishing renaissance that followed the "Meiji Restoration."

In January 1869, the following declaration was issued from the new capital of Tokyo:

> The Emperor of Japan announces to the sovereigns of all foreign countries and to their subjects [that permission has been granted to the Shogun Tokugawa Yoshinobu to return the governing power in accordance with his own request]. We shall henceforward exercise supreme authority in all the internal and external affairs of the country. Consequently the title of Emperor must be substituted for that of Taikun [the shogun], in which the treaties have been made. Officers are being appointed by us to the conduct of foreign affairs. It is desirable that the representatives of the treaty powers recognize this announcement.

Shortly thereafter, the young emperor (he would rule for forty-five years) boarded a Japanese naval vessel for the first time, and the next day gave instructions for studies to see how Japan's navy could be strengthened. There followed an extraordinary, indeed miraculous urge to join the modern world. It is true that upon so proud a nation Perry's brutal forced entry would leave a scar that time could never efface. However, for the foreseeable future, Japan, abandoning all national pride, would learn and copy from the West in every possible direction. In 1868 the young emperor himself spelled out in a charter oath his determination that "knowledge shall be sought for all over the world and thus shall be strengthened the foundation of the imperial polity."

A new, compulsory education scheme would create fifty-four thousand primary schools—or one for roughly every six hundred inhabitants; this would eventually lead to the Japanese becoming the most highly literate people in Asia. Within one generation, Japan subjected itself to an astonishing industrial revolution, one designed to catch up with two centuries of Western progress. The mantra for Japanese industry and learning became henceforth, unashamedly, and in general successfully, "copy, improve, and innovate."

Invariably, these developments acquired a military, and naval, flavor. Popular was the slogan "Rich country, strong army." Envoys scoured Europe for the best models to emulate. Initially it was the France of Louis-Napoleon, which was reckoned to have the world's finest army, just returned from the Crimean War. French shipyards were filled with orders for Japanese naval vessels. Then came the sudden debacle of 1870, and France's humiliation at Sedan at the hands of Bismarck and Moltke. Prussian military advisers were soon on their way to Meiji Tokyo, and British shipyards would shortly receive orders for the emperor's new model navy. Naval officers, like the future admiral Heihachiro Togo, were sent to Plymouth to study and train. Togo spent seven years with the Royal Navy, returning aboard one of the three new warships delivered to Japan.

Hand in hand with this martial renaissance went thoughts of imperial expansion. As little Japan watched the landgrab for Africa, led by the forces of another small island power, Britain, and in which even a newcomer like the upstart Kaiser could claim outposts as far away as the northern coast of China, an element of "me-too-ism" crept in. Predictably, this would bring Japan into conflict with the two neighboring imperial colossi, first China, then Russia.

At that time, Japan could cite demographic excuses for becoming expansionist. Its population in 1873 totaled nearly thirty-five million, roughly the same as Great Britain's. (By 1904, it would approach forty-seven million.) Its people were crammed into islands roughly the same size as the British Isles, but much of their mountainous terrain was as resistant to the plough as the Grampians of Scotland. And Japan had no outlets for expansion as did much-envied Britain in India, Canada, Australia, and the newly acquired colonies in Africa. The Japanese looked across the hundred-mile Straits of Tsushima to the fertile fields of Korea, and beyond them to the huge empty plains of Manchuria, rich in the mineral wealth, timber, and raw materials that their country lacked. Both were vassal provinces of China's groggy Manchu empire.

In 1894 Japan went to war with China. The nominal excuse was the assassination, and quartering, in Shanghai of a pro-Japanese Korean revolutionary, Kim Ok-gyun. But in fact the Sino-Japanese War was one of more or less straightforward colonial acquisition, and it gave the young, resurgent Japan an opportunity to flex its muscles and demonstrate its newly acquired martial hardware. The war lasted only eight months. Japan won virtually every round—especially at sea, where its modernized fleet (eight of whose ships were of British origin, to three French and two Japanese-built warships) sank eight out of China's ten warships, while the Japanese army inflicted disproportionately brutal losses upon the sluggish Chinese ground forces. Most of the Japanese dead had succumbed to disease. Japan ended up in possession of Taiwan (Formosa) and with a degree of control over a Korea removed from fealty to Peking. For the first time in over two thousand years, dominance in East Asia shifted from China to Japan. In China the humiliating loss sparked an unprecedented revolt against the tottering Manchu dynasty, eventually leading to the Sun Yat-sen revolution. In the harsh Treaty of Shimonoseki that followed the war, China also had to pay war reparations of two hundred million silver Kuping taels. According to the Chinese scholar Jin Xide, the Manchu government paid a huge sum of silver to Japan for both the reparations of war and war trophies, equal to about 6.4 times the total Japanese government's revenue.

In addition—and it was a vital concession—China agreed to cede the strategically important Liaotung (or Liaodong) Peninsula, with its key naval base that later became known as Port Arthur (now Lüshun, or Lüshunkou), a superb and well-protected natural harbor on the peninsula's extreme southern tip.* Among other things, the cession was an imposing sign of how far Japan had come since the Meiji Restoration, within only one generation. The European powers, consisting of Russia, France, and Germany, all with their own colonial interests and their own enclaves on the coast of China, however, were shocked by the draconian terms exacted by this upstart minor nation. They were also incensed by the Japanese troops' brutal massacre of Chinese soldiers and civilians in conquered Port Arthur—a glimpse of the future. The powers expressed concern that in Japanese hands, the port would present a "constant menace" to the Chinese capital, which lay less than three hundred miles to the northwest.

Russia, France, and Germany (Britain, keen to sell warships to Japan, was conspicuously absent from this imposing lineup) intervened to demand a revision of the Shimonoseki terms. Under the threat of war against overwhelming force (all Russian ships in Japanese ports received orders to be ready to sail at twenty-four hours' notice, in preparation for hostilities), Japan, though allowed to hang on to Tai-

*Port Arthur was named after a humble Royal Navy lieutenant, William C. Arthur, who during the Second Opium War in 1860 had towed his crippled frigate into the harbor of what was then an unfortified fishing village.

wan, was forced to cede Port Arthur back to China—which promptly leased it to Russia for a period of twenty-five years. In return, Japan was granted a further bribe of thirty million taels. It would be enough to reequip its navy and start up its own munitions industries instead of having to depend on European imports: all in time to prepare its war machine for the next round, which would come in ten years' time.

However, not for the first or last time, the proud Japanese deeply resented this interference and "disrespectful" condescension by the powers. In Tokyo the hand of the nationalists and expansionists was notably strengthened. The significance of Port Arthur's proximity to Peking, as an assurance against any future threat from the traditional foe, had certainly not escaped the strategically minded Japanese either. But not only that; the port was an ideal base from which to control Korea, or to launch any future venture into Manchuria just across the Yalu River. And, what was more, Port Arthur, sheltered within the embrace of the Yellow Sea, offered the most northerly ice-free port on the mainland.

This is what now brought Russia on to the scene. As of the turn of the century, tsarist Russia was both the largest and the most aggressively imperialist nation on the globe. Its advance across Siberia to the Pacific had begun with Yermak Timofeyevich, a freebooter Cossack, as early as the late sixteenth century, at the time of Ivan the Terrible. Yermak was eventually murdered by the exploited Tartars, and it was not until 1860 that a firm foothold was established on the Pacific, where a modest town was built at Vladivostok, opposite the northern Japanese island of Hokkaido. But the trouble with Vladivostok, though it was an excellent port capable of holding the whole of the Russian Pacific Fleet, was that it was iced in for three to four months of the year. In contrast, Port Arthur, protected by the almost landlocked Yellow Sea, was not—hence the Russians' passionate interest in acquiring it as a base, equaled only by the Japanese determination that they should not. Apart from its winter limitations, Vladivostok was at the end of the world's longest, most attenuated, and most tenuous railway line, the Trans-Siberian, single-tracked from Moscow for seven thousand miles. In the event of war in the Far East, the railway could be logistically disastrous: railway trucks had to be ferried for sixty miles across the vastness of Lake Baikal before performing a great loop around the north of Chinese Manchuria.

In 1896, a year after the Sino-Japanese War had ended, Russia wheedled rights from the battered Chinese to build a spur line straight across Manchuria, via Harbin, a new, Russian-built town, to Vladivostok. This line, the Chinese Eastern Railway, shortened the route by about four hundred miles. Only two years later, the Russians started to push through another spur line, the South Manchuria Railway, from Harbin through to Port Arthur—thus connecting the port city by rail to Moscow. Under the dynamic lead of Sergey Witte, the son of a German Baltic baron, a railways expert, and the tsar's minister of finance, work on the railways went ahead at phenomenal speed, unprecedented in shambling tsarist Russia—although Witte himself was said to have been strongly opposed to Russia's commitment to constructing a naval base at Port Arthur.

An anxious Tokyo seethed with rage over the Russians' rail enterprises. This was understandably so, as it looked as if the Russians, deploying a substantial armed force to protect the railway workings, were pulling off a de facto takeover of Manchuria, and then, with the second line—the South Manchuria Railway—entrenching themselves permanently in Port Arthur. Toward the end of 1903, Japan prepared for war against the Russian colossus. Most observers in the West thought it would never happen, and that this upstart Asian David could not possibly prevail against Saint Petersburg's imperial Goliath. Certainly the tsar's generals thought so. But, at the beginning of February 1904, the Japanese chief of staff, Field Marshal Iwao Oyama, told the emperor that it was essential for Japan to strike first if it wished to avoid the loss of its "birthrights" in Korea and Manchuria.

CHAPTER 2

Port Arthur

ON FEBRUARY 6, 1904, Vice Admiral Heihachiro Togo, aged fifty-seven, commander in chief of Japan's Combined Fleet, summoned his commanders aboard his flagship, the 15,000-ton *Mikasa*.* They were greeted by an unsheathed *sambo*, the short sword used traditionally by the samurai for the rite of seppuku, or hara-kiri, lying on the table in his cabin. It was a clear signal that war had been decided upon. Togo told his officers with great solemnity, "We sail tomorrow, and our enemy flies the Russian flag." He then proceeded to hand out detailed orders. Aboard his British-built flagship, Togo led his force of six battleships (all of them also British-built), ten cruisers, thirty destroyers, and forty torpedo boats out of Sasebo naval base on the southern tip of Japan. They headed into the Yellow Sea in the direction of Port Arthur. Before the fleet sailed, Japan's efficient spy networks on the mainland had supplied Togo with every detail of the location of the Russian fleet anchored safely in Port Arthur; they continued to alert him each time there was a change of berth.

Meanwhile another naval force escorting transports that carried three thousand army troops was directed toward the Korean port of Chemulpo (later called Inchon). There they were to create a bridge-

**Mikasa*, built by Vickers in 1900, with four twelve-inch guns, was then one of the most powerful battleships afloat, the last of the pre-*Dreadnought* era. The Russians had nothing comparable.

head from which Field Marshal Oyama's ground forces could threaten Port Arthur from the rear and advance northward into Manchuria, with the objective of preempting the arrival of Russian land reinforcements via the Trans-Siberian rail networks. Timing was everything, and neither time nor numbers were on the side of the attacker. The assault had to be carried out before the ice melted around Vladivostok, which would allow the Russian fleet iced in there to sortie into the Yellow Sea to reinforce the fleet at Port Arthur. And certainly it had to be concluded before the new Russian railways across Manchuria could come into play.

Togo dispatched ten fast destroyers (again, several of them British-built) streaking ahead of his main force to deliver a surprise torpedo attack on the Russian fleet insouciantly anchored in the security of Port Arthur. The first Japanese torpedoes, in what was in effect the earliest successful torpedo attack in the history of naval warfare, were fired just before midnight on February 8. As no declaration of war had yet been received, Saint Petersburg, as well as the Russian fleet, were taken completely by surprise. Here was a new style of diplomacy and warfare. Certainly it took the chancelleries of Europe by surprise; none had believed that the Japanese pygmy would really dare attack the Russian giant—least of all the giant itself.

At least it might have furnished a warning for the US Navy at Pearl Harbor a few decades later. The similarities are remarkable (the main difference being that on December 7, 1941, Admiral Yamamoto's force dropped their torpedoes and bombs from carrier-borne aircraft). In both incidents the element of surprise was achieved in that war had not been declared and diplomatic negotiations were still under way. As Commander Fuchida, the dashing leader of the attacking squadrons at Pearl Harbor, exclaimed in amazement: "Have these Americans never heard of Port Arthur?" But, to the dismay of us historians, our leaders tend not to read history, or, worse, they pay no attention to its lessons. Yet it does seem extraordinary that, in 1941 with tensions rising between the United States and Japan, Franklin Delano Roosevelt did not seem aware of the dangerous precedents; after all, he himself had been assistant secretary of the navy in the First World War, only a decade after the attack on Port Arthur, and was well read in history.

What of the Japanese commander on this momentous night in February 1904? Heihachiro Togo was born in 1848, the son of a samurai living in the city of Kagushima in feudal, pre-Meiji Japan. It was an area from which came the notorious Satsuma samurai, as a breed of dedicated militarists perhaps comparable to the Prussians in history. The Satsuma were a clique that stuck closely together. Togo received his first taste of battle at the age of fifteen when the Royal Navy bombarded the city in a bout of gunboat justice for the killing of a Briton the previous year. The following year he and two brothers enlisted in the daimyo's navy and fought in the last of pre-Meiji civil wars, serving in a paddle-wheel warship. Despite his early blooding by the British, Togo went with a group of other Japanese students to England in 1871 as an apprentice officer. He thought London a strange city and was amazed by the "abundance of meat," but found his own cadet rations "inadequate": "I swallowed my small rations in a moment. I formed the habit of dipping my bread in my tea and eating a great deal of it, to the surprise of my English comrades." His British comrades, ignorant of the difference between Asiatic peoples, called him "Johnny Chinaman." Occasionally such affronts would end in blows; nevertheless, Togo graduated second in the class.

In 1875, Togo went around the world as an ordinary seaman on the British training ship *Hampshire*, staying seventy days at sea without a port call until reaching Melbourne, and eating only salted meat and ship's biscuits. By the time he returned to Britain, he had sailed thirty thousand miles. Young Togo then suffered a strange illness that severely threatened his eyesight, but he was cured by Harley Street ophthalmologists after suffering a great deal of discomfort, a development that left him with an additional sense of gratitude to Britain. He subsequently went to the Royal Naval Academy in Portsmouth, followed by the Royal Naval College, Greenwich. While in Britain, he was able to inspect the construction of the *Fuso*, one of the warships being built for Japan at the Isle of Dogs. He returned to Japan in 1878 aboard another British-built ship, the *Hiei*. During his time in Britain he was described as a slender, pale-faced lad with conspicuously big lips, who showed "no promise of distinction but exhibited no weakness that might bar it." His captain reported: "Togo was an excellent fellow. He was not what you would call brilliant, but a great plodder, slow to learn, but very sure when he had learnt; and he wanted to learn everything! . . . one of the best sailors the *Worcester* has ever turned out."

At the beginning of the Sino-Japanese War in 1894, Togo, now a captain of the cruiser *Naniwa*, sank a British transport ship, the *Kowshing*, that had been chartered by the Chinese to transport troops. The sinking almost caused a diplomatic incident between Japan and Great Britain, but it was finally recognized by British jurists as being in conformity with international law. It also gave rise to unpleasant reports of what today might amount to war-crimes charges committed by Togo, involving the killing of the Chinese crew while they were struggling in the water. But all through history the Japanese would be known for their contemptuously brutal treatment of the Chinese. After the end of the Sino-Japanese War, Togo was successively commandant of the Naval War College (Japan) and commander of the Sasebo Naval College; then, in 1903, he was given command of the Standing Fleet. Because he was relatively unknown in imperial circles, Togo's appointment caused some surprise. The minister of the navy, Admiral Count Gonnohyoe Yamamoto, had to explain it to Emperor Meiji: "Because Togo is a man of good fortune." Napoleon, with his insistence that his generals be "lucky," would doubtless have understood.

Contemporaries describe Togo as being "temperamentally incapable of lifting a finger to gain the slightest preferment for himself," and found him to be "modest and unassuming." Yet he could write in the privacy of his diaries (in English), "I am firmly convinced that I am the reincarnation of Horatio Nelson." This was before his Tsushima triumph. As of 1904, he was a short, stocky man with a stubbly King George V beard and a thin graying moustache. He had a wide forehead and enormous ears, but was only five feet three inches in height. The war correspondent and artist H. C. Seppings Wright described him as having "a kindly face . . . marked by lines of care . . . the eyes are brilliant and black, like those of all Japanese, and a slight pucker at the corners suggests humour, a small drooping nose shades a pursed-up mouth with the under lip slightly protruding. He has a large head which is a good shape and shows strongly defined bumps, and the hair is thin and worn very short."

On the night of February 8, 1904, rather than a Nelson, Togo possibly bore a closer resemblance to a more cautious British admiral, John Jellicoe. Of Jellicoe it was said by Winston Churchill that, in 1914, he was the only leader who could lose the war in an afternoon. So could Togo. As well as time, numbers were against Japan, which had a population of only 46.5 million against Russia's 130 million, a population that, in theory, could provide limitless reserves. Russia could mobilize 1.1 million men, plus 2.4 million reservists, to Japan's standing army of only about 380,000 (of whom only a fraction were instantly available in the Far East). By 1904, Russia's Pacific Squadron boasted seven battleships, seven cruisers, twenty-five destroyers, and twenty-seven smaller ships. Japan's total fleet then numbered six battleships, ten cruisers, forty destroyers, and forty smaller craft, and these ships—almost all British-built—were superior in quality and speed to their Russian counterparts. Yet in a time of war Japan could count on no replacements; its shipyards had no facilities for building capital ships and were capable of only limited repairs. The Russians could draw, eventually, on substantial reinforcements from their home fleet, which included a class of five powerful battleships, though neither Port Arthur nor iced-in Vladivostok had any dry dock for substantial repairs. Both in 1905 and in the 1940s, the Japanese navy's admirals were to show themselves habitually and by tradition more cautious than the army's generals. So too now in the surprise attack of February 8, 1904: Admiral Togo could not afford to lose a single one of his precious capital ships.

On acquiring the lease to Port Arthur, Tsar Nicholas had exulted in his diary: "At last an ice-free port!" That was about the best that could be said of it. In February 1904, in the words of one writer, Port Arthur resembled Chicago at its worst: "rough, immoral, and vulgar. A Kansas barnyard, one visitor called it. . . . From the bay, Port Arthur looked picturesque, but it was derelict and decaying, a collection of jerry-built stone houses, temporary warehouses, and equally temporary administrative buildings."

An American gold miner from the Klondike gold rush of 1897 might equally have found similarities with the Alaskan port of entry of Skagway. Port Arthur was a town of whores and booze; one hotel, the Effimiev, possessed only twenty-four beds in a dark, dingy, and filthy shedlike building that was described by a contemporary traveler as the meanest ever provided for civilized men. But it was also a garrison town populated by some eighteen thousand soldiers and twenty thousand sailors and dockworkers, and housing Chinese coolies in miserable conditions—upon all of whom the Russians had imposed their own peculiar brand of squalor.

Yet above the slum, the tsar's personal envoy, the sixty-year-old Vice Admiral Yevgeny Alexeyev, presided in his grand viceregal lodge. An incompetent and venal figure, Alexeyev almost certainly owed his rank to the fact that he was the illegitimate son of Tsar Alexander II, "The Reformer," the grandfather of Nicholas II. Between 1895 and 1897 he commanded the Russian Pacific Fleet, initially at Vladivostok, until it moved to Port Arthur in 1897. Next, Alexeyev became governor of the newly leased territory, and then commander in chief of the Pacific Fleet again. He was regarded in Tokyo as a hard-liner most hostile to Japan, so his appointment in August 1893 as plenipotentiary viceroy, in charge of both military and civil functions over the whole of Russian-held territory in the Far East, had been taken as another omen that negotiations would bring no favorable concessions. It was the first time an admiral had been appointed to such high office, and Nicholas's able ex-minister Sergey Witte considered Alexeyev "a nincompoop," and certainly "not an army man" because he could not even ride a horse. At least he had every reason to know the area extremely well; in consequence, he personally could be held responsible for much of the extreme dilapidation of the Port Arthur defenses.

Most serious was the failure of the Russian defenders under the indolent Alexeyev to keep the key harbor of Port Arthur, which had been so coveted in Saint Petersburg, properly dredged. In consequence, all seven of the Russian capital ships on the night of the attack were actually riding at anchor outside the port, in the roadstead. There, even though supposedly protected with torpedo nets, in their three parallel lines they were much more vulnerable to attack.

In Port Arthur the night of Togo's surprise attack, all was merry insouciance. The town and the Russian warships were brightly lit up. The Russian commander, the fifty-eight-year-old Vice Admiral Oskar Stark, described as "an affable old sea dog, growing weary and absent-minded," had an English nurse tending his children while he was celebrating a birthday party for his wife on the deck of his flagship, the *Petropavlovsk*. Most of the fleet's officers were aboard. On shore "a gay atmosphere pervaded the streets, one could see people staggering tipsy along the streets." Many ships' crews were prowling the bars on Pushkin Street, or carousing in brothels such as one called the American Legation.

Togo had ordered his destroyers to approach without lights and with engines damped down so that their presence would not be betrayed by sparks belching out of funnels. Nevertheless some witnesses saw the approach of the Japanese destroyers coming in as a long line of moving lights, like fireflies. Ten minutes before midnight, the historic attack commenced. At first, guests at the admiral's party thought the exploding Japanese torpedoes were being fired in honor of the birthday. A witness wrote:

> The band was playing loudly, and the most honorary guest, Admiral Alekseyev, the viceroy, glided along the parquet with the heroine of the day with surprising grace for his rather obese figure. When the dance reached its apogee . . . the windowpanes suddenly shook from the thunder of the cannonade. Everyone applauded, surprised by such a timely salute, and the overall excitement increased. The ball continued to the accompaniment of the orchestra and artillery fire.

After the first torpedoes had struck, wild firing came from the Russian ships, many shooting at each other in the confusion. The Japanese attack was swiftly over. There were also moments of panic among the inexperienced crews aboard the attacking destroyers, but nothing compared to the chaos that reigned both on the Russian ships and in the town. On the other hand, there were people like the London *Times* correspondent, asleep on a ship in the harbor, who knew nothing of the attack until the morning. When the Russians, licking their wounds, came to count their losses, they found that two battleships, the *Retvizan* and the *Tsesarevich*, the most powerful in the fleet, had been holed and were sitting on the bottom in shallow water, and the cruiser *Pallada* had been sunk with coal bunkers afire. *Retvizan* had a hole in her side measuring 220 square feet. Yet this was only two out of seven, and both battleships would be refloated. Considering the risks involved of an undeclared war against mighty Russia, the results looked pretty poor.

That night, although his fleet had practiced the operation several times, Togo committed several cardinal errors through inexperience and excessive caution. In the destroyer attack he failed to have at least one ship provided with a radio to inform him of the results. Faulty reconnaissance resulted in four of his precious battleships being hit, suffering sixty casualties, despite the wildness of the Russian shooting. At the same time, full of exaggerated fears of the danger presented by the Russian batteries, he withdrew after making only a single pass. In fact, such was the shock of the attack that the batteries were largely unprepared for action. Captain William Pakenham, the fearless and perceptive British naval observer who was at Togo's side all through the war, excuses this on the grounds that, to Togo, it would have been "incomprehensible that he would find everything in such a state of unpreparedness." One biographer of the admiral, Edwin A. Falk, reckons that had Togo deployed all his available force with full vigor, "the Port Arthur squadron could have been wiped out" and "the control of the sea virtually would have been won before the war had commenced."

Criticism should perhaps be levied against Togo as the first twentieth-century naval commander to have the advantage of striking such a surprise blow. In contrast to the Americans at Pearl Harbor, Togo was faced by a somnolent, incompetent, and lethargic enemy, inert in its supreme self-confidence. Admiral Yamamoto has been criticized for his excessive caution in December 1941 for not following up on his first success by destroying the Pearl Harbor oil depots and ground installations. He had also failed to catch the aircraft carriers commanded by US admiral William F. Halsey Jr., which were then out at sea and were to prove the decisive weapons in the war to follow. But, by comparison, Togo in his caution, if not his pusillanimity, seems to deserve more severe castigation. He had better intelligence than his successor; he knew the position of every single Russian capital ship—with both the factor of complete surprise and weapon superiority should he not have been able to sink the lot? Did he in fact prove, in this operation, to be a less successful admiral than Yamamoto? On the other hand, he was a green commander, in charge of new weaponry, in a novel form of offensive warfare.

So, Togo's surprise preemptive attack proved at best a qualified success; indeed, it could easily have ended in disaster for Japan. The Western press did not on the whole react with disfavor to Japan's ungentlemanly attack. The London *Times* robustly endorsed it as "stunning" and "masculine." "Our ally," it declared, "put her navy in motion with a promptness and courage that exhorted the admiration of the world . . . this [was a] dashing and courageous exploit." After all, the Russian fleet had been inviting attack by standing in the outer roadstead. These were hardly the words that would be used on the "day of infamy" thirty-seven years later; but the brutal, autocratic country of the tsar of 1904 was hardly the flavor of the day with liberal Britain. Suddenly Europeans began to revise their opinions of the "little yellow men." Only France, dependent upon its own alliance with Russia, was swift to criticize.

Upon the Russian navy crews at Port Arthur the psychological impact was grave; little effort was made to venture out and confront the attackers. When Russian warships were sent out, on February 11, disaster struck. A minelayer, *Yenisey*, and a destroyer, *Boyarin*, both ran onto mines that *Yenisey* herself had just laid, and two Russian destroyers were damaged in a collision. *Yenisey* sank at once; *Boyarin* stayed afloat, despite two risibly incompetent Russian attempts to torpedo her to keep her out of Japanese hands. The following day a boarding team was sent to salvage the wardroom silver, and she sank during a storm that night. The captain was duly court-martialed and deprived of command for a year.

The fortunes of war swung back and forth. The Russians mined the entrance to Port Arthur to prevent the Japanese from entering again, and the Japanese lay down mines to deny to the Russian warships access to the open sea. At Chemulpo, in contrast, where the

Japanese launched a diversionary attack on the Korean coast, a heroic fight was fought by *Varyag*, a modern US-built cruiser commanded by Captain Vsevolod Rudnev. Refusing Japanese calls to surrender while anchored in the port, Rudnev put out to sea to face overwhelming odds, declaring to his crew, "Let us put our trust in God, and go bravely into battle for the Tsar and for the Motherland—Hurrah!" He then sailed toward the enemy, his ship's band playing the national anthem, "God Save the Tsar." *Varyag* fired 1,105 rounds from her various guns, though not one found a target. Hammered into a wreck, with Rudnev wounded and with thirty-one of her crew dead and another ninety-one seriously wounded, *Varyag* was scuttled. Reaching Odessa, Captain Rudnev, as the only hero of the war so far, was accorded a triumphant welcome with parades and banquets.

Meanwhile, the Japanese Twelfth Division had successfully established its bridgehead in Korea and was thrusting northward for the lower reaches of the Yalu River at Antung—names that would become ominously familiar to the UN armies during the Korean War.

On February 24, Togo, reckoning he could no longer count on sinking the Russian capital ships outright, endeavored to block them in Port Arthur. To this end he sent five old merchant ships full of ballast to be sunk in the harbor's mouth. This meant more or less imposing a death sentence on the ships' skeleton crews, all volunteers. Apparently two thousand applications were submitted, some of them "earnestly written in blood." But the attempt was thwarted by heavy shellfire from the Russian shore batteries, and from the two-year-old battleship *Retvizan*, aground in the mud but able to bring her powerful twelve-inch guns to bear. All five of the blockships were destroyed before they could be positioned. A second abortive attempt to block the harbor was made on March 27; and Togo attempted yet a third on April 27. Russian morale had been vastly improved in the meantime, though only temporarily, by the arrival of a new admiral to replace the disgraced Stark.

Vice Admiral Stepan Makarov was the ablest of all Russia's admirals, one among few. Upon his appointment, there were cries of "Makarov is coming! Makarov is coming!" across the Pacific Squadron. To the demoralized sailors he was "the closest thing to a savior." Ironically, given the Japanese choice of weapons in their surprise attack, Makarov—a prolific inventor—had been one of the progenitors of the deadly torpedo boat. In January 1878, during the Russo-Turkish War, he had been the first in the world to launch torpedoes from a "fastboat," against an Ottoman navy ship, *Intibah.* He was highly decorated for his services during that war.

It is hard to tell from portraits what manner of man was Makarov, gentle or harsh, intelligent or insensitive. In the style of the times, his face was covered with an impenetrable fuzz—a vast moustache and bifurcated beard, features not dissimilar to those of his exact contemporary and fellow admiral Alfred von Tirpitz of Germany (also an exponent of torpedo development). Makarov was a huge man, "with self-confidence and a prestige second to none." He soon became known, affectionately, to every officer and seaman in the fleet as "Old Beardy." After Russia took over Port Arthur in March 1898, Makarov wrote to the naval minister, "The fall of Port Arthur would be a terrible blow for our position in the Far East. Port Arthur should be made impregnable." He also suggested that forty torpedo boats should be disassembled and sent to Port Arthur by the Trans-Siberian Railway. Unfortunately for Russia, few such measures were ever pursued before the outbreak of hostilities in 1904.

Makarov arrived to take over command in Port Arthur on March 7. Unlike any other Russian leader in the war, he was aggressive and innovative, and he reportedly had a rare ability to "inspire confidence in his subordinates." He at once stepped up activity in the Russian squadrons, as well as improved the defenses of Port Arthur. Shipwrights were set to work repairing the ships damaged in the initial Japanese torpedo attack. Because Port Arthur lacked a dry dock big enough to take the battleships *Retvizan* and *Tsesarevich*, Makarov had cofferdams laboriously constructed, large wooden boxes that fit against the ships' hulls. Once in place, the cofferdams were pumped out, giving the shipwrights a dry working space right down to the bottom of the ship. The valuable battleships were refloated; in terms of numbers, Togo's surprise attack had been all but annulled.

Up till then the Russian fleet had done virtually nothing but exist on the defensive, as a fleet in being. A riddle made the rounds: "Why do Russian squadrons need two admirals? Answer: One to take the ships to sea, and one to order them back." Under Makarov's leadership, his ships now put to sea nearly every day, and so were constantly on the move, ensuring that they were never taken by surprise outside the protection of Port Arthur's shore batteries. Unlike his predecessor, Makarov energetically sought engagement with the Japanese, and when Japanese cruisers bombarded Port Arthur from the Yellow Sea in March, his cruisers returned fire with such efficacy that the Japanese ships were forced to withdraw. Rather than risking a major engagement, Makarov's objective was to build up technical competence and morale. Eventually his aim was to break out of the cul-de-sac of the Yellow Sea and sail with his fleet to Vladivostok, the more effectively to harry Togo's fleet, disrupt Japanese lines of communication, and strike at the Japanese homeland.

In the middle of one briefing, Makarov was told that the destroyer *Steregushchy* was pinned down in action with Japanese destroyers outside the port. Without saying a word, Makarov left the conference. Minutes later, the small cruiser *Novik* was leading the ships out of harbor and flying the admiral's flag. Cheers went up from those watching; and even though it was too late to save *Steregushchy*, Makarov had decisively demonstrated that he intended to lead from the front.

But his new command troubled him. "Oh, to know what to do!" he wrote. "Truly our men are in need of everything. They do not know how to walk in the night. Mismanaged and confused, they continue to elbow each other near Port Arthur. Incapable of identifying themselves they hesitate to return [to port] for fear of being mistaken for the Japanese. Complete misfortune!" Nonetheless, that same month, when the Japanese made their second attempt to seal the port's entrance with blockships, Makarov's cruisers vigorously attacked the escorting Japanese warships and put them to flight.

Then, on April 13, providence deserted the reanimated Russians. A Russian destroyer, *Strasny*, returning from patrol, tried to reenter Port Arthur but was intercepted by Japanese destroyers. An engagement began between the opposing destroyers, and Makarov immediately sent the cruiser *Bayan* to assist *Strasny*, while he himself led a force of three battleships, four cruisers, and some destroyers into the Yellow Sea to seek battle with the surrounding enemy warships. The Japanese warships withdrew out to sea, with Makarov in pursuit. With the arrival of enemy heavy units upon the scene, Makarov, now out of range of gunfire support from the shore, turned about and headed back to Port Arthur. The Japanese fleet did not pursue. The dashing Makarov had fallen into a trap. The previous night a Japanese minelayer, *Koryu Maru*, had laid a field outside the harbor. The Japanese destroyers covering *Koryu Maru* were observed by Makarov, but the minelayer herself went undetected. As his flagship, the battleship *Petropavlovsk*, moved closer to the harbor's entrance, she struck one or two mines in the newly laid field; moments later, her forward magazines exploded, with such violence that the entire forward turret was thrown into the air.

Captain Vladimir Semenov aboard the cruiser *Diana* described the scene graphically. He heard a violent explosion, then a second.

> Suddenly cries of horror arose: "The *Petropavlovsk*! The *Petropavlovsk*! Dreading the worst, I rushed to the side. I saw a huge cloud of brown smoke. In this cloud I saw the ship's fore-

> mast. It was slanting, helpless, not as if it was falling, but as if it were suspended in the air. To the left of this cloud I saw the battleship's stern. It looked, as always, as if the awful happenings in the fore-part were none of its concern. A third explosion! White steam now began to mix with the brown cloud. The boilers had burst! Suddenly the stern of the battleship rose straight in the air. . . .

Petropavlovsk sank within two minutes; some observers said they saw the admiral kneeling in prayer on the bridge as the ship went down. Word spread that Japanese submarines were nearby; the Russian ships began firing wildly in all directions. Witnessing the sinking from the Japanese battleship *Asahi*, William Pakenham recorded that "the silence of the Japanese fleet was broken by an involuntary burst of cheering." Destroyers sent to search the site of the disaster found only fifty-eight survivors from *Petropavlovsk*; over 680 officers and men had been lost with the ship, among them most of Makarov's staff, and his close friend the famous war artist Vasily Vereshchagin, noted for his paintings of Napoleon's retreat from Moscow. Of Admiral Stepan Makarov himself, the searchers found only his overcoat. Throughout Russia, "a grief-stricken nation," there was somber mourning for the death of the admiral. Leo Tolstoy criticized the regime bitterly for sending innocent sailors to their deaths.

The Japanese fleet paid handsome tribute to a fallen enemy hero, such as would not be repeated in a later, less chivalrous, and more brutal Pacific War. Togo himself ordered all flags to be flown at half mast, and a day's mourning for the Russian. In Port Arthur morale plummeted. The fleet once more sat locked in the harbor. Outside in the Yellow Sea, Togo's cruisers prowled like cats waiting for a mouse to venture out from its hole, ready to pounce. Togo further enhanced the growing competence of his young navy by sending a task force north to bombard Vladivostok and the Russian units trapped inside.

He then mined its approaches, thus further separating the two Russian naval units. It was all good training for the Japanese, while Russia's frustrated sailors gained nothing but further humiliation. At least for the time being, Japan had assumed command of the seas. In Saint Petersburg, contempt for the "little yellow monkeys," and the belief that their threats would never amount to anything, was replaced by a certain sense of alarm. The dispatch of army reinforcements, led by General Aleksey Kuropatkin, down the long tube of the Trans-Siberian was urged on, and there was discussion about sending naval reinforcements from the Baltic; but, as was normal in tsarist Russia, nothing happened except vacillation.

Supposing, however, that Makarov had not gone down with *Petropavlovsk*, might the Russo-Japanese War have taken a different turn? One of the more recent historians of the war, J. N. Westwood, remarks: "Although Makarov was subsequently credited with a genius which he may not have possessed, he was certainly Togo's equal, and perhaps more." One thing was certain: tsarist Russia could not field anybody better. Nevertheless, in Tokyo celebration of the sinking of *Petropavlovsk* and the elimination of a redoubtable foe was muted. It was not only the wise heads of the antiwar lobby who recognized that the longer the war went on, so did the odds against a Japanese victory lengthen.

CHAPTER 3

Into Manchuria

THE FOCUS OF the war now moved from sea to land as Japan's armies pushed northward from Chemulpo into North Korea and the bleak landscape of southern Manchuria. Now may be a good place to compare the quality of the opposing armed forces, and notably their most essential ingredients: the fighting soldiers and sailors. A Russian epigram of the time about postings for newly commissioned officers ran, "Drinkers to the fleet, dimwits to the infantry," and there was more than an element of truth to it. Because of the iced-in conditions, Russian Baltic Sea crews spent six months ashore, boozing; the same applied to forces at Vladivostok. Little time could be spent on gunnery. The contrast between the shipboard living conditions of the officers and those of the crewmen was far worse than in other navies. The officers often showed little concern about distributing rotten meat to the crews. Consequently, morale in general was poor. By comparison, Japanese crews lived well. Under their British instructors, they spent much more time out at sea, and trained harder. Most Japanese sailors had been raised on or near the coast; many were fishermen. On the other hand, few Russian sailors had ever seen the sea prior to conscription; few were literate, in contrast to the Japanese. And these days of steam and big-gun technologies placed heavy demands on the ill-educated.

On land, the Russian muzhik or peasant would fight sullenly and die stoically for his tsar. In contrast, under the ethos of the samurai, the Japanese fighting man felt that to die for his emperor was the highest achievable honor. The *British Official History of the Russo-Japanese War*, for instance, speaks admiringly of Japanese values, not least "that wonderful spirit of self-sacrifice which animates the heart of every man—soldier or sailor—and makes him feel it a privilege to give his life, if by that means the welfare of the nation may be advanced." In marked contrast was the Russian serviceman, for whom "the dull surroundings of his village home deaden his imagination and produce a stolid nature which, even after frequent defeats, is usually proof against sudden panic or disorganization." Out in the field Russian infantry commanders would be averse to deploying in "extended order"—because it increased the difficulties of command. This was a mentality that was to cost the Russians dearly as the land battles developed. Thus infantrymen on both sides in 1904–5 were ready to accept heavier losses than would their Western equivalents—that is, after 1914.

As regards intelligence and spying, the Japanese also had a distinct advantage. Both sides treated the indigenous Koreans and Chinese with contemptuous brutality; pictures of the times show Russian soldiers hanging recalcitrant locals from the trestles of railway bridges, and the Japanese lopping off their heads with ceremonial swords. But by and large the Japanese came out ahead; they paid their spies better. Apart from his first serious misjudgment, Togo was consistently well informed about the moves of the Russian fleet. To the intense dismay of the Russians, the Japanese could also boast of a highly well-placed and informed intelligence and disinformation network inside Europe, and indeed inside Russia itself. Paranoia about an infestation of spies swiftly spread among the Russian forces, uncovering networks everywhere across Manchuria, and provoking excessive alarm.

By coincidence, both opposing generals in 1904, Field Marshal Iwao Oyama and General Aleksey Kuropatkin, had been present at the historic Battle of Sedan in September 1870 to witness the cataclysmic, dynasty-changing defeat of France by Helmuth von Moltke's

Prussians. Born in Kagoshima, and, like Togo, of Satsuma samurai blood, Oyama had instantly fallen for the martial virtues demonstrated by the Prussians, though he was only twenty-eight at Sedan. He later became one of the founders of the Imperial Japanese Army. (He was also, incidentally, the first recorded Japanese customer for Louis Vuitton, having purchased some luggage during his stay in France.) Under his influence, and true to the "copy and improve" principles of the Meijis, the Japanese army had then selected Prussians to advise and train its forces. The Prussian general Jakob Meckel had taught the Japanese how to establish an efficient general staff. The Prussian doctrines of discipline, *Gehorsamkeit*, and opportunism, the *Auftragssystem*, together with the bark of the bullying *Feldwebel*, were not unsuited to the ethics of the samurai and Bushido. For better or for worse, they would remain at the heart of the Japanese army for decades to come. In consequence, one observer with the Japanese army in 1904–5, the future general Sir Ian Hamilton, later of Gallipoli fame, regarded them as "the world's best infantry."

During the Sino-Japanese War of 1894–95, Oyama, then a general, had led the seizure of Port Arthur—for which triumph he was made a marquis and, three years later, a field marshal. In 1904, as a venerable sixty-two-year-old, he was nominated commander in chief of the newly formed Manchuria Army, though he had been among those who had opposed the war. His operative chief of staff was a tough nut called General Gentaro Kodama, who had worked with General Meckel and had then gone on to be military attaché in Berlin.* An anti-Russian hawk, Kodama had also served as a hard-line minister of the army under Prime Minister Ito Hirobumi. Not unlike

*Although his period in Japan (1885–88) was relatively short, Meckel had a considerable impact on the development of the Japanese military. He is credited with having introduced Clausewitz's military theories and the Prussian concept of war games (*Kriegsspiel*) to the Japanese. His reforms are also considered to have played a role in Japan's overwhelming victory over China in the 1894–95 Sino-Japanese War.

the Ludendorff-Hindenburg relationship (the German command on the western front in the First World War), it was Kodama rather than Oyama who laid down the strategy of the whole land campaign of 1904–5.

The Russian general, the fifty-six-year-old Aleksey Kuropatkin, however, perhaps groomed in Tolstoyan tenets and convinced that all Germans were innately stupid, paid little heed to the lessons of the victors at Sedan. The son of a minor noble from Pskov Province, Kuropatkin has been described as "a bureaucrat's general." Appointed war minister in 1898, he made it his priority to improve conditions for both officers and men, as well as the quality of the officer corps. None of his reforms achieved much success. Although counted as an expansionist, a believer in railways, he opposed war with Japan, deeming that southern Manchuria was untenable by Russian arms. A trip in June 1903 to Japan, where he was impressed by the state of the Imperial Army, strengthened these views. He arrived in Manchuria to take overall command of the tsarist land forces in March that year, with something of a fatalist, if not defeatist, view. If he had a strategy, it was decidedly defensive. Arriving at the chief Russian base at Liaoyang on the crucial railway line, just south of Mukden, Kuropatkin prepared to sit it out and wage a war of attrition. He would avoid any offensive action until the Trans-Siberian Railway had brought in decisive reinforcements in troops and matériel. But the Japanese were not going to give him the time he needed. His command would be marked by caution and indecision, which would lead to repeated Russian defeats.

Once Togo had been assured that the Russian navy, bottled up in both Port Arthur and Vladivostok, could not interfere, Kodama landed the rest of the Twelfth Division at Inchon on February 16. Despite the bitter winter weather, Kodama's troops swiftly occupied the Korean capital, Seoul, some twenty miles inland. Nine days later Japan imposed the status of a Japanese protectorate on Korea, though it was not officially annexed until some six years later. Thus was completed the work of Hideyoshi four centuries earlier. Korea would remain a colony of Japan, often savagely repressed, until 1945. Two further Japanese divisions landed between March 14 and 21, and were grouped together with the Twelfth Division into the Japanese First Army, under the command of General Tamemoto Kuroki. Without waiting for the end of the spring thaw, which made the primitive Korean roads all but impassable, Kuroki thrust north toward the lower reaches of the Yalu River, that strategic and historic river that over its five-hundred-mile length demarcated the frontier of Korea and Manchuria. Early European visitors to Korea remarked that the country resembled "a sea in a heavy gale" because of the many successive mountain ranges that crisscrossed the peninsula. Some 80 percent of northern Korea is composed of mountains and uplands, bitterly cold in winter, divided by deep and narrow valleys. In the west, however, the coastal plains are wide and flat. It was to these that Kuroki's invading force wisely stuck in 1904.

Brushing aside a feeble Cossack force that arrived from Manchuria, weary and travel-stained, Kuroki, with some 42,500 men backed by 10,000 Korean coolies and porters, reached the Yalu in mid-April. Strategically and tactically, the march was not an overwhelming task for the Japanese army; it followed exactly the same route the Japanese had a decade previously in the Sino-Japanese War. Many of its senior commanders, such as General Maresuke Nogi (who had commanded the First Infantry Brigade), had accompanied that expedition and remembered it well. With admirable intelligence and deception measures (the Russians were constantly hamstrung by an almost total lack of intelligence about the enemy), Kuroki built a fake timber bridge conspicuously across the wide lower Yalu, to attract Russian fire; meanwhile, ten further pontoons were laid out upstream. Sketches of the time show troops crossing in sampans towed by steam launches, or lashed together and propelled by oars over the stern. Fabricated by Krupp of Essen, Kuroki's 4.7-inch howitzers, heavier than anything in the Russian armory, were used with devastating effect on the defenders. The Russian commander, General Mikhail Zasulich, was urged by his staff to pull back to a more defensible position. However, also ignoring General Kuropatkin's orders for a phased withdrawal, Zasulich stubbornly refused, even sending a telegram to the tsar in Saint Petersburg informing him that victory was at hand. On May 1, 1904, remnants of the Russian Eastern Detachment either surrendered or escaped northward.

Thus the Battle of the Yalu River, the first major land campaign of the war, ended in victory for the Japanese. It had cost them 1,000 dead and wounded, while the Russians suffered some 2,700 casualties overall, including about 500 killed, and the loss of twenty-one out of twenty-four field guns. The defeat of the Russian Eastern Detachment finally removed any remaining notion that the Japanese would be an easy enemy, that the war would be short, or that Russia would be the overwhelming victor. The major outcome was that the attackers were now across the main river obstacle separating them from the back door to Port Arthur.

Four days later, General Nogi's Third Japanese Army, ninety thousand strong, landed at the base of the narrow Liaotung Peninsula, on whose tip sat Port Arthur. These troops swung left, toward Port Arthur, while the rest of the Japanese forces, the Fourth, Second, and First Armies, under Generals Nozu, Oku, and Kuroki, combined to push northward up the railway to deny Kuropatkin any chance of coming down to relieve Port Arthur. The outlook for Port Arthur began to look distinctly grim.

Then, in the way of the unpredictable fortunes of war, disaster suddenly hit Togo's blockading squadron. Two of the Imperial Japanese Navy's capital ships, the battleships *Hatsuse* and *Yashima* (built in Newcastle), struck mines laid by the Russians in the open approaches to Port Arthur. To lay mines in neutral waters was a serious offense against the existing rules of war and was completely unexpected by Togo. Both ships sank, and, to add to the catastrophe, the cruiser *Yoshino* was sunk after a collision in the fog. At a stroke, the Imperial Japanese Navy had lost one-third of its capital force; the Russians now boasted a notable superiority in Pacific waters—at least on paper. With their habitual flair for secrecy, the Japanese managed to hide the sinking of one of the battleships till after the fall of Port Arthur. But, had Russia still had a Makarov at hand, the pendulum of war might easily have been made to swing in a decisive reversal of fortune against Japan. With a superiority now of six to four, an audacious commander could well have broken the blockade and chased off the surviving capital ships. In this event, the huge land force that the Japanese navy was now protecting and supplying in Korea and Manchuria would surely have been left out to dry on its long and exposed lifeline. One cannot imagine Allied navies in the Second World War missing such an opportunity, let alone the Japanese themselves.

But there was no Makarov now available to Russian arms. In his place was the inept and slothful viceroy of the Far East, Admiral Yevgeny Alexeyev, a pet of the tsar, whose first reaction to Togo's surprise attack on Port Arthur had been that it was "impossible." Though he had been a leading hawk before war started, early May saw the viceroy and his entourage prudently pulling out of Port Arthur, to Ying Kou in Manchuria, before his escape route could be cut.

The absconding Alexeyev then placed his junior, Vice Admiral Wilgelm Vitgeft, born in Odessa but—like so many of the tsar's leading sailors and soldiers—of German descent, to take over command of the Port Arthur fleet. Vitgeft is described acidly by Togo's biographer Edwin A. Falk as "qualified by temperament and long hours of duty ashore to entwine the propellers with red tape and inertia," and, when Vitgeft finally flew his flag aboard the *Tsesarevich*, as "a worried desk-sailor very much at sea." Vitgeft consistently rejected the now safely land-bound viceroy's urgings to break out of port. His cautiously modest view was to await the arrival of the Baltic Fleet from Saint Petersburg and the stream of reinforcements coming in at a snail's pace along the railway. But the hawkish viceroy, so far misguided in all his actions, constantly prodded Vitgeft to break out of the Japanese headlock and hit their depleted fleet. Had it been Makarov, Alexeyev might have been right. But Vitgeft sat tight, with his six capital ships to Togo's four.

The situation on land was becoming critical, turning decisively against the Russians. On May 26, at the Battle of Nanshan, General Oku's Second Army had cut the narrow, vital isthmus that connects Port Arthur with the Manchurian hinterland. Other Japanese army forces were rolling the rest of the Russian army northward, up the railway toward Mukden, before Kuropatkin's reinforcements could intervene effectively. As Port Arthur's predicament became grimmer, its isolation definitive, Alexeyev at last appealed to the tsar to force Vitgeft to move. Faced with imperial censure, Vitgeft, flying his flag in the *Tsesarevich*, led his fleet from the harbor early on August 10, 1904. Almost apologetically he signaled to his ships: "The Tsar has ordered our squadron to sail to Vladivostok." It was make or break, Vladivostok or bust, and smash impertinent Togo on the way.

The Russian fleet consisted of an impressive array of battleships: *Tsesarevich*, *Retvizan* (both refloated since their sinking in February), *Pobeda*, *Peresvet*, *Sevastopol*, and *Poltava*; the cruisers *Askold*, *Diana*, *Novik*, and *Pallada*; and fourteen destroyers, plus a similar number of torpedo boats. But under prevailing wartime law, as combatants they could be prevented from recoaling anywhere else once out of Port Arthur. From his advanced outpost in the Elliott Islands, conveniently tucked just off the Liaotung Peninsula, Togo spotted the large belch of black smoke as Vitgeft stoked up his boilers. He was on to him in a trice. His aim was to head off the Russian in a Nelsonian crossing of the T before Vitgeft could leave the Yellow Sea.

There would now follow a historic encounter, the first ever between the all-steel, all-steam behemoth capital ships that would set the pattern of naval warfare for the next two generations. Since that first encounter of the *Monitor* and *Merrimack* half a century previously, progress in naval weaponry had been meteoric. The combination of electric power and hydraulics enabled battleship designers to install hugely heavy armored turrets that could traverse and elevate mammoth guns at speed. But they had so far been untried; to this point, the war had been fought largely with the newer deadly weapons—the mine and the torpedo. Thus neither of the combatant navies, Russian or Japanese, could have had any conception of the damage that would be inflicted by the monster cannon of 305 mm (twelve-inch) caliber, which had a range of 15,000 yards, more than eight miles, firing a massive shell weighing 850 pounds (each gun alone weighed nearly fifty tons). Nor could they conceive of the punishment that these high-explosive projectiles could inflict on their own ships. And how would the new cannon perform after repeated firing? The two fleets were steaming into an unknown—and rather frightening—world.*

Shortly after midday there was some tentative firing between the two fleets at a long range of over eight miles. No hits were scored. Relying on his ships' superior speed, Togo adopted an elliptical course southeastward, moving out of sight of the Russians, but dogging them all the way with his light and fast cruisers. He came close to losing them. Many Russian sailors must have breathed sighs of relief as the Japanese capital ships disappeared. Perhaps Togo was not going to risk a further encounter against a superior foe after all? Meanwhile, down in the hellish boiler rooms the sweat-drenched stokers in both fleets were being pressed to supreme efforts: the one to escape to open sea, the other to prevent that escape. Indeed, at that point the engage-

*At the last fleet engagement, Trafalgar, almost exactly one hundred years previously, Nelson's biggest gun could fire projectiles weighing only sixty-eight pounds, and ranges were generally seldom more than a murderous two hundred yards.

ment was very much a "stoker's battle." Under such pressure, ships of both fleets faced the danger of a boiler-burst, which would scald the crews to an instant death, or a major engine breakdown. *Poltava* was already limping. For Togo, whose fleet could maintain sixteen knots to Vitgeft's fourteen (with luck), the race was also one against the day's remaining light. With only a few hours to go, the Russians might well elude him in the dark, or turn unexpectedly and maul his outnumbered capital ships. As the two Russian fleets combined at Vladivostok substantially outnumbered Togo's, the outcome would determine the whole course of the war.

By midafternoon, Togo was overhauling, and his leading ships were getting in range. The risk of T-crossing is that the lead ship—which is likely to have the admiral himself on board—initially takes the heaviest damage. Contrary to expectation, Russian gunnery on August 10, 1904, turned out to be extremely accurate, if not superior to the enemy's, especially at long range. So reported Britain's fearless observer with the Japanese, Captain Pakenham.* Togo's *Mikasa*, in the van, took several hits. Splinters from one heavy shell exploding on the mainmast narrowly missed Togo, who refused to take cover in his ship's armored conning tower, but insisted he remain on the exposed compass bridge.† When the battle was over, *Mikasa* had been hit twenty times—more than was sustained by all but one of Vitgeft's ships. Worse still, no fewer than five out of sixteen of Togo's twelve-inch battleship guns had been knocked out by their own armament in the course of the action, shells exploding in super-hot breeches with disastrous consequences. It appears that either the excessively sensitive

*Ironically for a nation that in later years would excel in the manufacture of precision optical equipment, Japan equipped its capital ships at this time with British Barr & Stroud range finders, whose serviceable range was only 6,600 yards.

†Observation from warship conning towers was through narrow, protective slits, not unlike that of a closed-down army tank. Hence commanders in both world wars tended to eschew them as being too limiting to their overall view of the battle scene.

Japanese-designed fuses or their own explosive, called *shimose,* tended to "cook off" inside the gun tube. This could easily have cost Japan the battle—and the war. As it was, however, by an extraordinary twist of fortune it may have caused the opposite.

Shortly before dusk on August 10, Vitgeft's flagship, *Tsesarevich,* was hit by two Japanese twelve-inch shells in quick succession. Instead of being armor-piercing projectiles designed to penetrate the vitals of a ship and then explode, with often mortal consequences, these seem to have been filled with *shimose* explosive and detonated by supersensitive fuses that were touched off by a ship's superstructure, or even by the rigging. One blew to pieces the unfortunate Admiral Vitgeft, of whom only one leg was ever found. He had so far that day performed better than expected, but—given his limitations as a commander—the consequences of his demise were far surpassed by those of the second shell, which burst on the slit of the conning tower, its blast killing or knocking out all its occupants. Jammed by dead bodies, the wheel of *Tsesarevich* remained immovably stuck at "hard-over," with the result that the Russian flagship performed a tight circle, nearly ramming two other battleships and disrupting the line she had been leading. Wrote Edwin A. Falk: "No one remained above decks to report below what had happened or to send a signal. So far as the rest of the fleet could tell, there was nothing particularly significant about those two puffs of smoke on the *Czesarevich* at a time when the entire line was sustaining hits." The rigging that carried her semaphore flags had been shredded, as well as her nerve center destroyed, so none of *Tsesarevich*'s subordinate capital ships could perceive what had happened; and, presuming that Vitgeft was pursuing some deliberate course of his own, like good, obedient Russians they attempted to follow his strange course. Chaos swiftly ensued. The Russian line was broken, its separate units "moving independently like parts of a severed earthworm," in Falk's vivid phrase, with Togo pouring in devastating fire at close range from his secondary armament as well as his heavy weapons. "Had the shells pierced the deck and burst in the bowels of the ship," concludes Falk, "the *Czesarevich* might have been destroyed but the result would have been less decisive than it was. The other Russian ships would have observed and understood what had happened."

As it was, the coherence of the Russian squadron was shattered, and—as dusk gathered—so it scattered. Five battleships, a cruiser, and nine destroyers escaped and crept back to Port Arthur; however, the damaged *Tsesarevich* and three escorting destroyers sailed to Tsingtao, where the Germans were obliged to intern them. The cruiser *Askold* and a destroyer reached Shanghai, where the Chinese interned them, while the *Diana* ran all the way to Saigon and French internment. The extraordinary thing about this ferocious seven-hour engagement, the first gunfire battle between iron leviathans, was that not a single ship was sunk. Although the damage to the *Tsesarevich* proved far and away the most critical in the battle, Togo's *Mikasa*, hit twenty times, had had her aft twelve-inch gun turret knocked out, and she had suffered 125 casualties—by far the heaviest of the day. Togo has been criticized for squandering his initial position of advantage, and young hawks in the Imperial Japanese Navy may have grumbled about his failure to move in at dusk and sink what remained of Vitgeft's ships. But he didn't need to. It had been a clear-cut strategic victory for Japan. Without a ship being lost, August 10 was a defining day in the war. Thenceforth, Russia's proud First Pacific Squadron, like Admiral Scheer's German fleet after Jutland in 1916, would not venture out of port again. It would sit in Port Arthur until scuttled, or sunk by the Japanese land batteries that were steadily moving in. There would be no question now of these ships linking up with the Vladivostok forces. Dismounted, their heavy guns were moved ashore as entrenched batteries to face the approaching Japanese armies.

With the Russians now effectively neutralized in Asian waters, the focus of the war once more returned to the land. On August 16, following the defeat of Admiral Vitgeft, Togo and Nogi jointly signed a letter to the Russians, inviting them to surrender in the hopes of avoiding a "useless sacrifice on a large scale of lives and property." The invitation was refused. The slaughter on land now began to escalate. The veteran General Nogi landed with the ninety-thousand-strong Third Army at the root of the Liaotung Peninsula after the Russians had been pushed back into it at the key Battle of Nanshan. He had done it all before in the Sino-Japanese War of ten years earlier, and the terrain was familiar to him. His express role was to advance down the peninsula and lay direct siege to Port Arthur. Meanwhile, the First, Second, and Fourth Japanese Armies, under Generals Kuroki, Oku, and Nozu, respectively, would carry on pushing the Russians northward up the railway toward Mukden, to interdict Kuropatkin's Trans-Siberian reinforcements arriving to relieve Port Arthur. For the Japanese, perpetually on the offensive, it was, as always, a race against time before superior Russian numbers could swing the tide of battle. As Russian resistance mounted, with the well-known stoicism of the Russian soldier fighting on the defensive, so Japanese casualties soared. Two of Nogi's own sons were to die in the battle.* Tactical overreliance on the use of infantry attacking frontally, as taught by the Prussian Meckel, was later considered to have contributed to the large number of Japanese casualties in the subsequent actions.

New levels in the intensity of infantry warfare, where the Maxim machine gun and field artillery dominated, were experienced daily. For the first time in the history of war, the Russian defenders would introduce the barbed wire entanglement to stop the massed Japanese infantry. They would also develop the field telephone, though

*Attempting to console Nogi's bereaved wife on the loss of her two sons, Togo received a reminder of samurai stoicism: neither of her children, she said, had "brought disgrace on the true name of a soldier and this gave her sufficient consolation."

its value would become somewhat limited when frostbitten Russian troops cut down the telegraph poles for firewood. There were many such harbingers of the First World War, just ten years away, and Western observers like General Sir Ian Hamilton were there alongside the Japanese commander to take note. The lessons didn't seem to sink in; Hamilton would lose his reputation, as well as many thousands of British and Anzac lives, at Gallipoli only eleven years later under not dissimilar conditions. Yet the Russo-Japanese War on land was rightly to become regarded as the first "modern" war—if only military minds in the West could have taken serious note of it.

The Japanese commanders discovered that infantry attacks in massed formation, as taught to them by their Prussian instructors, were prohibitively costly. The Russians in turn realized that their much vaunted Cossack cavalry were useless. Six massive eleven-inch Krupp howitzers, destined for Nogi's siege forces, were sunk at sea by Russian raiders creeping out of Vladivostok, and their loss would be felt bitterly by the attacking Japanese, as they were forced to reduce one fiercely contested strongpoint after another. All through September, and into October and November, as the autumn rains turned the battlefield into a bog, like Passchendaele in the First World War, Nogi's men clawed their way forward, in relentless and sometimes suicidal frontal assaults. To replace the losses, an extra sixteen thousand reinforcements had to be shipped in. But the defenders, now isolated, could count on neither reinforcements nor food supplies.

Then, in October, disturbing news from Europe reached Tokyo. After endless vacillations, the tsar had finally dispatched the bulk of his home naval forces (the Baltic Fleet), imposingly renamed the Second Pacific Squadron, to the Far East to reinforce what remained of Makarov's and Vitgeft's First Pacific Squadron and relieve beleaguered Port Arthur. The new squadron would face an eighteen-thousand-mile world cruise to reach the battle zone; it was indeed late in the day, if not far too late. But, with his numerical superiority still at risk, the news that one of the major naval forces in Europe was on its way could not fail to cause Togo deep concern. If this Russian fleet could break his battle line and then seal off the four Japanese armies in Korea and southern Manchuria, the nation would face extinction. Nogi was instructed to press ahead with the seizure of Port Arthur at full speed, and regardless of cost. There was talk among the Tokyo "doves" about proffering peace terms, before the tide of battle truly turned.

The battle swayed back and forth for the possession of a modest height known as 203 Meter Hill, which dominated Port Arthur at less than three miles from the harbor. It had a clear view for almost point-blank fire on the surviving vessels in it. One by one these were sunk by Nogi's heavy siege guns. Rations in the beleaguered city began to run out, and horses and donkeys were eaten. Because of the shortage of vegetables and fruit, cases of scurvy proliferated, as did acrid recriminations between the defeated navy and the army that continued to fight. Of the Russian army commanders themselves, Lieutenant General Anatoly Stoessel—another German, described as an "uncouth-looking man with a stentorian voice"—countermanded every order given by his rather more effective subordinate, Konstantin Smirnov, and in turn fell out with the commander in chief, Kuropatkin. He was denounced by both as "not a fighting general." Those other handmaidens of the misery of siege warfare, dysentery and typhoid, began to take hold. The Japanese meanwhile suffered from shortages of rice, crucial to their rations.

Winter was beginning to blow in from the cruel uplands of Manchuria. Nogi moved in over four hundred guns; he had the advantage of bringing them up under the superb cover of fields of millet, which had been allowed to grow ten feet tall, and which the defenders had neglected to mow. Still Port Arthur held out. There were even reverses, as when the Japanese attempted to board the battleship *Reshitelny,* only to be repulsed and the Japanese commander chucked overboard. Morale sank on both sides; there were cases of Russian regiments firing on each other, and even reports of Japanese units, having suffered very heavy losses, refusing to leave their trenches. One Japanese regiment, thrown into the assault 1,800 men strong, emerged four days later with only 200 alive. Two attempts to seize 203 Meter Hill were beaten off at huge cost, and it was finally taken only in late November. Surging back and forth, with tunneling and countertunneling, the bitter fighting presented a foretaste of what the western front would be like a decade later.

So did the butcher's bill. In two days, the Japanese would lose another ten thousand men with little gain. Feelings against General Nogi became intense on account of his terrible casualty lists, as the Japanese had to storm uphill over open ground. There were proposals that he should be replaced. Then his second, and favorite, son was killed on the deadly hill. Officers were authorized to kill those men who, without proper reason, straggled or retreated. In a remarkable custom, instead of dog tags, Japanese officers would send home to families of the killed the dead soldier's Adam's apple.

On the Russian side, their best and bravest commander, Colonel Nikolai Tretyakov, put up a desperate defense comparable with that of the defenders of Stalingrad in 1942, using homemade grenades in hand-to-hand fighting. Frostbite, lice, and scurvy were as much the enemy as the Japanese. Terrible was the stench of the wounded—though, in the arctic cold, they mainly froze to death. By the end of November, every inch of the ground of 203 Meter Hill had been ploughed by the shells from the eleven-inch Krupp howitzers that Nogi had brought up, manhandled into position by teams of eight hundred soldiers; "rocks were ground to powder," reported the correspondent-cum-artist H. C. Seppings Wright. On November 30, the courageous defenders finally gave up. From the captured hill, Nogi's howitzers now completed the final destruction of the fleet at their mercy in the harbor below. The once noble Russian Pacific Squadron at last succumbed to shore fire and torpedo attacks. A total of 120 torpedoes were fired at *Sevastopol*; she survived, only to be scuttled ignominiously by her crew.* The *Retvizan*, sunk on February 8 but resuscitated by Makarov, was sunk again by the Japanese howitzers.† It was indeed a "messy end to the fleet that had once been the late Admiral Makarov's pride."

Meanwhile in the besieged town conditions had become intolerable. Hospitals had run out of soap, bedpans, and mattresses. Dysentery was universal. The Cossacks took to treating the Chinese inhabitants with unbridled brutality. Stoessel asked the tsar for permission to surrender. On January 2, 1905, the battered and starved port finally caved in. On occupying the port, Nogi was surprised to find large stores of food and ammunition, suggesting that Stoessel had surrendered prematurely. Now the Russian looters took over. Upon returning to Saint Petersburg, Stoessel was court-martialed and sentenced to death, though this was later commuted to ten years' imprisonment. Over 30,000 Russian soldiers and seamen surrendered their arms, and about 15,000 sick and wounded remained in the squalid hospitals. The siege had cost Japan 57,780 in killed and wounded; dysentery, beriberi, and other diseases brought the total to around 100,000—losses that this small nation could ill afford. The Russians were estimated to have lost at least 31,000 men.

Though his two sons had died in the fighting, Nogi's terms for the defeated were chivalrous—they would stand in marked contrast to the treatment of British and US troops surrendering to Japanese arms in 1942. Together with Togo, Nogi clambered to the top of Hill 203,

---

*Its remains still lie outside the entrance to Port Arthur.

†After the surrender of Port Arthur in January 1905, the Japanese raised and repaired her. She was commissioned in the Imperial Japanese Navy as *Hizen* in 1908. When the Japanese declared war on Germany in 1914, she was sent to reinforce the weak British squadron in British Columbia. She was disarmed in 1922 in accordance with the Washington Naval Treaty and sunk as a gunnery target in 1924.

still strewn with the dead. Nogi invoked the "spirits of the dead" to share his victory, begging the emperor for forgiveness for the excessive number of soldiers' lives he had lost. At the war's end, in contrition he asked the emperor's permission to commit seppuku by way of atonement. "Not while I'm alive," came the response. But by and large the fall of Port Arthur was greeted with jubilation and immense outbursts of patriotism. "Of all the news which reached Tokyo during the war," wrote the British correspondent Ellis Ashmead-Bartlett, "there was none which caused such intense joy." A new generation was coming along, too: a Lieutenant Hideki Tojo was celebrating his twentieth birthday at military college the day before the fall was announced. Three decades later he would lead Japan into its most disastrous war, and meet his end on an Allied gallows.

In the world outside, the fall of Port Arthur caused the credit of Russia to fall, and that of Japan to rise. In Saint Petersburg there were the beginnings of revolutionary grumblings; Lenin would write that the capitulation was "the prologue to the capitulation of Tsardom."

Notable among Japanese intelligence agents was the ex-military attaché in Saint Petersburg, Colonel Motojiro Akashi, who on the outbreak of war moved his quarters to Stockholm. From there he acquired an extraordinary field of contacts among Russian dissidents and revolutionaries bent on the downfall of the regime; these included Lenin, for whom Akashi would provide funds for the first issue of his newspaper, in January 1905. To the High Command in Tokyo, Akashi was reckoned to be worth ten divisions.

A short while later Akashi would be in London, arranging the shipment to Russia of arms for the revolution that would explode in 1905. In December a strike broke out at the important Putilov plant, a supplier of munitions to the armies in the Far East. Prominent among the workers' demands were calls that the war should stop. Such a strike was unprecedented in tsarist Russia. But it spread swiftly until eight hundred thousand workers were out.

Then, on Sunday, January 22, 1905, the real trouble began. A mysterious but charismatic Russian Orthodox priest, Father Georgy Gapon, led an entirely peaceable crowd of several thousands to the Winter Palace in Saint Petersburg to present petitions to the tsar (who in fact was at Tsarskoe Selo, "the tsar's village," about fifteen miles away). After warning shots, the panicky troops guarding the palace fired directly into the crowd, killing an estimated one thousand innocent demonstrators—men, women, and children. After "Bloody Sunday," the gauntlet was down. In February the tsar's uncle, Grand Duke Sergei Alexandrovich, would be assassinated. For tsarist Russia these were the first steps toward revolution; more would follow as further bad news came from the Pacific.

At Port Arthur, this first, most terrible battle of the twentieth century where so many thousands had died, the fighting did indeed seem like the ugliest harbinger of the horrors that would soon begin in Europe: at Verdun, on the Somme, in Galicia, on the Piave. But, taking place as it did in this far-off corner of the globe, only a few would see it, and even fewer would note its awful significance, its warnings of what heavy artillery, the Maxim machine gun, the bolt-action rifle, the land mine, barbed wire, and a determined defense could do to humanity.

Port Arthur having surrendered, to its north in Manchuria, along the railway lifeline to Mukden, Oyama's armies—totaling 128,000 men to the Russians' 158,000—hammered Kuropatkin's forces. Several times, Kuropatkin, now named supreme commander in place of a disgraced Yevgeny Alexeyev, launched counterattacks in the bitter Manchurian weather, conditions in which Russian troops traditionally excelled. At last major reinforcements were funneling in over the creaking railway system. But despite their growing numerical superiority, the Russians in Manchuria suffered from a variety of handicaps. Kuropatkin and his subordinate commanders were wedded to defensive doctrines (two of these subordinates, Generals Samsonov and Rennenkampf, would find themselves commanding armies in East Prussia at the beginning of the First

World War, with disastrous consequences). In contrast, Field Marshal Oyama instructed his generals to attack constantly—before Kuropatkin's reinforcements could reach the front. As all through the war, Japanese intelligence proved superior; Russian commanders arrived in southern Manchuria without any up-to-date maps of the area. The Russians' communications could hardly have been worse; not knowing their own positions, they frequently saw cases of friendly fire. Morale was little better, with reports of broken units raiding vodka stores. And transport remained chaotic, with engineless goods wagons sometimes being pushed by squads of sixteen coolies. Each Russian offensive was repulsed, but at an enormous cost to both sides. Japan was rapidly reaching the bottom of the barrel in terms of its manpower reserves. And yet, as Hitler would find in the Second World War, Russia's reserves seemed to be limitless.

The bitter winter conditions imposed a lull in the fighting. But by the end of February 1905, both sides were squaring up for the decisive battle, for the key center of Mukden. Acting with their customary aggression and vigor, the Japanese armies—now at their peak strength—struck to preempt the counterstroke that Kuropatkin had been so sluggishly preparing. Given the superiority he had assembled, especially in artillery, with a total of more than 1,400 pieces to the Japanese 924, his opponent Oyama had to move swiftly and adroitly. He chose a copybook "pin-and-hook" strategy of which Napoleon would have approved. Coming out of the hilly country to the east, General Kawamura's newly formed Army of Yalu, whose strength was unknown to Kuropatkin, attacked energetically to give the impression that this was the main assault. With the Russians pinned down in that direction, Nogi's army, released from the siege of Port Arthur and now a highly experienced, battle-toughened force, was "looping boldly" to the west. It would cut the railway nearly ten miles north of Mukden, and well to the rear of Kuropatkin's general headquarters.

An attempt by Kuropatkin to meet the challenge, by splitting off portions of General Alexander Kaulbar's Second Manchurian Army, was disrupted by a fierce blizzard on March 2. Panic spread, and demoralized Russian soldiers rushed to loot the vodka stores. Even so, the Japanese remained at risk, with Nogi's army not seeming strong enough to close the jaws of the trap. A more talented commander than Kuropatkin might have seen an opportunity there. But the opening was missed. On March 7, with the railway north of Mukden now cut, Kuropatkin telegraphed Saint Petersburg tersely: "I am surrounded." With that he cut his losses and abandoned Mukden.

It was a disorderly and disgraceful retreat. Much of Kuropatkin's munitions and transport was lost. Discipline collapsed, and Russian troops turned on their officers and drunken men roamed the streets. There were many self-inflicted injuries. The *British Official History* recounts: "There were carts in matchwood, wagons with their wheels in the air . . . every kind of baggage lay crushed by the horses' hooves or by the wheels of vehicles which had passed over them."

Kuropatkin was sacked by the tsar and replaced by another incompetent general, Nikolai Linievich. The army withdrew to a safe point two hundred miles north of Mukden. The whole of valuable southern Manchuria was lost to Russia. But it was not a rout; Oyama's forces were too exhausted to follow. They had lost more than a quarter of the troops committed to the battle. General Gentaro Kodama admitted, "It was never thought possible by us that we could surround the Russians and bring about a second Sedan." Indeed, had the Russians been capable of mounting a strong counteroffensive at that moment, Oyama's army would have been caught dangerously exposed.

The cold was crippling in its intensity, tougher on the unacclimatized Japanese than on Kuropatkin's Siberians. Losses had been horrendous on both sides, far higher than had been experienced in any one of Napoleon's battles, or for that matter even on killing fields like Gettysburg in the American Civil War.* At the Battle of Muk-

*Leipzig, in 1813, Napoleon's costliest battle, resulted in 38,000 French casualties to 54,000 of the Allies. Gettysburg cost a total of 51,000 for both sides.

den, in 1905, out of sight of most of the world, Japan lost 15,892 in dead and 59,612 wounded; Russia about 40,000 killed or missing and 49,000 wounded—equal to about one-third of each side. Many of the wounded would not survive. The Russians had also lost most of their heavy weaponry. Mukden was rightly judged to be the bloodiest battle in history up to 1905. By the end of the battle, both sides were ready for a settlement, for a negotiated peace. It marked the end of any serious fighting on land in the Russo-Japanese War. Although on paper Japan had been the winner on both sea and land all along, in fact it was the more in need of an end to the fighting. Having relied on a short war, it had lost too heavily, in terms both of blood and wealth. And Russia could always mount another offensive from the Trans-Siberian Railway. As it was, its mighty new fleet, the Second Pacific Squadron, was already nearing the end of its eighteen-thousand-mile expedition to avenge, or reconquer, Port Arthur.

But revolt was simmering in European Russia.

Meanwhile, in southern Manchuria, as the spring thaw brought an end to campaigning, both sides sat down and licked their wounds. For Japan, at least, Mukden and the territory captured in southern Manchuria, plus of course Port Arthur, would be valuable bargaining counters when it came to a peace settlement.

As both sides gasped for breath, the final and most dramatic act of the war, the battle with which it would forever be identified, remained to be played out, at sea once again.

CHAPTER 4

Odyssey, Followed by Iliad

WE NOW APPROACH the last, melodramatic act in the extraordinary saga of the Russo-Japanese War. The climactic battle, for which Russia's newly formed Second Pacific Squadron spent more than half a year voyaging eighteen thousand miles around the world, was to be fought and won within a span of roughly half an hour. As Admiral Togo reflected afterward, pondering the unevenhandedness of fate, was there not something gross in the fact that a great battleship could take several years to construct and train up, yet could be destroyed completely in a half hour? Thus, similarly, the destiny of nations.

The man the tsar chose to lead the Second Pacific Squadron on its epic voyage around the world to defeat Togo and recapture Port Arthur was vice admiral Zinovy Rozhestvensky. The tsar, as usual, was unerring in his ability to make the wrong choice—though, since the demise of Makarov, the Russian list of available candidates was not overwhelmingly large. The question is: Could any naval commander have done better than Rozhestvensky, on a voyage that was to prove an almost impossible task, all but a death sentence? In the struggle just to keep his crews together, maybe no one could have performed better than he.

Rozhestvensky was born the same year as Togo (1848) and had begun his career as a gunnery officer with the Baltic Fleet. During the Russo-Turkish War of 1877–78, as a lieutenant, his torpedo boat, equipped with spar torpedoes,* had been caught up in the nets deployed to protect the enemy ships, and his formation had had to withdraw without success. In 1891 he began three years as naval attaché in London, where he gained a somewhat grudging respect for the Royal Navy, then regarded as without rival, and in 1894 he served under Makarov in Russia's Mediterranean Squadron. But promotion was slow. Aged fifty in 1897, Rozhestvensky was a mere captain in command of the gunnery school of the Baltic Fleet, his retirement looming. It was at this point that he caught the eye of the tsar. The occasion was a review in honor of the tsar's cousin, Kaiser Wilhelm II of Germany, on July 24, 1902. At a naval review off Reval (now Tallinn in Estonia), Captain Rozhestvensky laid on a brilliant display of gunnery that greatly impressed the naval-minded kaiser. The rapidity and accuracy of the big gun salvos, the result of weeks of fiercely tough training, were outstanding. The kaiser lavished praise on the Russian commander—"I wish I had such splendid admirals as your Rozhestvensky"—and as he departed he offered a typically arrogant compliment: "The Admiral of the Atlantic bids farewell to the Admiral of the Pacific."

Rozhestvensky was a martinet with an explosive temper. He was nicknamed, very much in private, "Mad Dog." In a rage he was given to throwing his binoculars overboard. A huge figure of a man with an impressive bearing, his trim salt-and-pepper beard made him resemble the tsar's other eminent cousin, King George V—except for the arching, very Russian eyebrows.

As a consequence of his performance and the kaiser's praise at the 1902 maneuvers, Rozhestvensky caught the tsar's attention. From then on his promotion was meteoric. That same year he emerged as a vice admiral, and chief of the naval staff. Proposing a plan to send

---

*Spar torpedoes, attached to the bowsprit, required the ramming of the enemy ship, and thus close combat, which in turn necessitated a great deal of courage.

most of the Baltic Fleet around the world to reinforce the embattled ships in the Far East, he suddenly found himself in command of it. Was he up to the job? This unprecedented task would be fateful and ultimately fatal. One historian regarded him as "the right man for the job, for it would take an iron-fisted commander to sail an untested fleet of brand-new battleships [for some of the new *Borodino* class, this voyage was their shakedown cruise] and new untrained sailors on the longest coal-powered battleship fleet voyage in recorded history." Another historian, rather less charitably, regarded Rozhestvensky as "a screaming imbecile, promoted well past his abilities . . . suffering from a near mental breakdown" even before the battle he was being sent to face. Most of his officers were reputed to "tremble before him like thieves before the constable. He treats them worse than a bad master treats servants."

Typical both of the vacillations of the last Emperor of All the Russias and of the poor state of his warships, week by week, month by month, as the odds lengthened against the survival of Port Arthur, the departure of the newly rechristened Second Pacific Squadron, originally set for June 1904, was delayed. Part of the trouble, despite the presence of the new battleships, was the poor state of the other ships, some of which had seen service during the Russo-Turkish War of a generation earlier; the delays also reflected the haphazard way in which sailors were recruited by an essentially nonmaritime nation. Ships' complements were largely made up of a mix of Germans, Poles, Balts, Jews, and Finns; many had no naval training whatsoever, and there were also a substantial number who brought vehement revolutionary sympathies, if not tracts, on board with them. Driving himself as rigorously as his crews, Rozhestvensky had to work eighteen hours a day, often going without sleep for three nights in a row, to get his ships ready, roaring at his staff and terrifying the commissary merchants who came on board. As well as ammunition, shells, torpedoes, and mines by the ton, food for his twelve thousand seamen had to be provided, along with clothing for both tropical and Arctic conditions. Herds of cattle were driven on board to provide meat, and dried biscuits and vegetables, and vodka and champagne for the officers' wardrooms, were loaded.

Most critical in the era before the oil-fired boiler were the vast quantities of fuel, which had to be stocked—thousands of tons of coal. With no Russian base en route, the whole issue of coaling would remain Rozhestvensky's most constant headache throughout the seven-month cruise. No other nation in history had undertaken such a voyage.

The admiral raised his flag on the new, Russian-built *Knyaz Suvorov*, named after the great general of the eighteenth century, a *Borodino*-class juggernaut that displaced over fifteen thousand tons. One of the most modern warships afloat, built in a Russian shipyard to a French design, she had been in commission only a matter of weeks and had had few comprehensive sea tests. For no very clear reason her twin funnels were conspicuously painted a bright canary yellow. Eventually the fleet's departure date would slip to October 9. It was late in the day, if not too late. Russia's best admiral, Makarov, had been killed, likewise his successor Vitgeft; and all attempts to break through the Japanese blockade of Port Arthur had failed. By land the Japanese army divisions were relentlessly moving in on the port. Even the crucial hub of Mukden was under threat. The day Rozhestvensky finally sailed, with cheerful military bands and gay bunting along the waterfront, both the tsar and tsaritsa came to watch the fleet lift anchor. They could not help being impressed by the bearing and "rock-like authority" of the newly appointed commander in chief. He would go far to restore Russia's battered reputation in the Far East. He also sailed with the additional blessing of the kaiser: in the web of alliances and counteralliances that was building up in the years before 1914, Wilhelm II calculated that either Japan, the naval ally of his archenemy Britain, would be humiliated, its British-built warships shown to be defective; or Russia, the best friend in Europe of his traditional foe France, would be trounced. It must have looked like a no-lose situation.

As the historian Richard Hough notes, "Never before in the whole annals of naval warfare had a Commander-in-Chief sailed forth on such a long, such a dangerous and difficult mission with such an ill-assorted selection of vessels." The mighty flotilla, fifty ships strong, was built around four of the navy's latest battleships and three older ones. The voyage began inauspiciously. Unable to count on resupply from neutral or unfriendly ports anywhere along its route, all Rozhestvensky's ships started off critically overloaded. The battleship *Oryol*, laden down to her gunwales with victuals, ammunition, and coal, ran aground on a sandbank. It was a humiliating and unhappy omen. A few days later, the *Suvorov* also ran aground at Libau (then a leading port in the Russian Baltic, now in Latvia). To superstitious Russians these were bad signs. Among the farewell speechmaking at the send-off, one of the commanders, Captain Bukhvostok of the *Alexander III*, struck a grimly prophetic note: "We know why we are going to sea. We also know that Russia is not a sea power and that the public funds spent on ship construction have been wasted. You wish us victory, but there will be no victory. . . . But we will know how to die and we shall never surrender."

Meanwhile, Rozhestvensky alarmed his officers by ordering them to show extra vigilance, as there were reports of Japanese torpedo boats lurking in the North Sea disguised as trawlers. Any suspicious ships were to be sunk on sight. This was all good "disinformation" work by Japan's astute naval attaché, Motojiro Akashi, who had been relocated to Stockholm on the outbreak of hostilities. Akashi's false intelligence was picked up by Russian reports out of Copenhagen, which appeared to substantiate rumors about the torpedo boats being at large. Consequently, tension, if not alarm, aboard the flotilla was high: noted one officer, Chief Engineer Politovsky of the *Suvorov*,

"We must be on our guard . . . panic prevails on board. Everyone examines the sea intently."

On the night of October 21–22, 1904, as the great fleet plunged on toward the English Channel, Rozhestvensky's worst fears crystallized just off the Dogger Bank. It was a typical North Sea night for that time of year: black, spectral, with limited visibility. While still in Danish waters the repair ship *Kamchatka*, lost in fog, suddenly radioed that she was being attacked by about eight torpedo boats, coming from all directions. Rozhestvensky's nervous commanders promptly stood to, and in short order the sea was ablaze with searchlights, and with heavy guns firing in all directions. The *Oryol* alone fired off some five hundred shells. Great spouts of water were sent up close to the *Suvorov*. The enemy appeared to be attacking from every point of the compass. In the melee, one Russian cruiser, the *Aurora* (later to become famous in the 1917 revolution), was hit by friendly fire; her chaplain lost a hand and subsequently died of blood poisoning—the fleet's first casualty. But the engagement seemed to be going Russia's way. Three enemy ships had been hit; one was listing heavily. Then the dreadful penny dropped: the Japanese "torpedo boats" were in fact small, hundred-ton trawlers of the British Gamecock fleet, out on their business from Hull—and, of course, unarmed. Aboard the *Crane* most of the crew had sustained wounds, the captain had been decapitated and the trawler was sinking. Many of the other British trawlers had also been hit.

When the news reached England, outrage swept the country. A fierce editorial in *The Times* denounced the Russians unsparingly: it was "almost inconceivable that any men calling themselves seamen, however frightened they might be, could spend twenty minutes bombarding a fleet of fishing boats without discovering the nature of their target . . . butchering poor fishermen with the guns of a great fleet without endeavoring to rescue the victims of their unpardonable mistake."

Anger spread electrically. Trafalgar Square was filled with protesting crowds; the Russian ambassador was booed as he left his embassy; there were deputations to Downing Street and to Parliament calling for action. To many in liberal Britain, the authoritarian tsarist regime was anathema anyway, and few wanted the pending extension of the Entente Cordiale to embrace Russia too.* Japan was Britain's ally. Moreover, the offense was exacerbated by the fact that it had taken place on Trafalgar Day, of all days, the centenary of which was to be celebrated in exactly a year's time. The king, Edward VII, deeply angered, summoned the foreign secretary, Lord Lansdowne, to the palace. He donated two hundred guineas for the victims, and penciled in the margin of the report to Parliament: "A most dastardly outrage." Even the German press, then normally favorable to the Russians, damned Rozhestvensky, questioning whether his fleet should be permitted to be at sea at all. In Saint Petersburg the tsar wrote privately to his mother that the English "are very angry and near to boiling point. They are even said to be getting their fleet ready for action. Yesterday I sent a telegram to Uncle Bertie, expressing my regret, but I did not apologise. . . . I do not think the English will have the cheek to go further than to indulge in threats."

Yet by October 25, it did indeed look as if Britain was on the brink of war. "The mind of the government, like the mind of the nation, is made up," thundered *The Times*. The Home Fleet, commanded in the North Sea and with guns loaded and at the ready, was dispatched to Portland. Fast cruisers were alerted to shadow the Russians out of the Channel. By the evening of the twenty-sixth, there were twenty-eight battleships, with steam up, ready to destroy Rozhestvensky's squadron at one word from the Admiralty. Gibraltar was put on a war

*The Triple Entente, the alliance between France, Britain, and Russia, followed in 1907 and is thought by many historians to have been among the contributory causes of the First World War.

footing, and British warships were recalled from foreign ports. The Russians were closely shadowed, almost escorted, all the way to the coast of Africa.

Until the Russians arrived at Vigo Bay, in northwest Spain, Rozhestvensky seems to have been sublimely unaware of the storm he had precipitated. There, among other British journalists, Edgar Wallace, the detective writer, then a reporter for the *Daily Mail*, interviewed members of the Russian fleet. They were still convinced they had genuinely come under attack by Japanese torpedo boats disguised as trawlers. Then the penny really did drop for the Russians, who discovered the truth. The crisis ebbed. Amends were made. Rozhestvensky admitted his errors; Russia agreed to pay a compensation of £65,000 to the families of the dead and injured fishermen. It had been a close-run thing. But being shadowed all the way by the cruisers of Admiral Lord Charles Beresford "as if they were prizes of war" caused the already jumpy nerves of Rozhestvensky's sailors to fray still further, as the Russians could not help noting superior seamanship beyond anything of which they were capable. A dashing figure who commanded the Channel Fleet and was normally accompanied by a very British bulldog, Beresford teased the Russians by darting ahead and crossing their bows, then falling astern to steam past at full speed in perfect line-ahead formation, with humiliating closeness to Rozhestvensky's plodding fleet. Rozhestvensky himself observed gloomily to his subordinate, now-commander Vladimir Semenov: "Those are a real squadron. Those are seamen. Oh, if only we—" He broke off in midsentence and hurried down a companion ladder. But British anger over the Dogger Bank incident would pursue Rozhestvensky for the rest of his voyage.

The most immediate problem that the crisis imposed on the Russian commander was in the urgent matter of refueling. The recoaling of warships was considered to be one of the most distasteful tasks, and it was often dangerous. Until the fuel for capital ships was replaced by oil in the First World War, they had to be refueled at regular intervals with vast quantities of coal. Every spare man and officer was brought in to do the coaling; stokers often died of heart failure in the heat. During the coaling the whole ship was filled with dust, which penetrated even the cabins and wardrooms of the officers. At a single recoaling, a battleship could take on over two thousand tons of coal, which could add to the danger of the ship listing and ultimately overturning. Any ships caught when coaling in a time of war were sitting targets.

Before Rozhestvensky's departure, Saint Petersburg had skillfully done a deal with the Hamburg-Amerika line to provide coalers the whole length of the fleet's extraordinary voyage. With Kaiser Wilhelm anxious to distract European powers from his own naval designs, the Hamburg firm would faithfully fulfill its coaling contract. But the problem lay with the ports on the route; with Russia being a belligerent power, they could lawfully refuse Rozhestvensky access for the lengthy and tricky process that coaling entailed. And coaling in the open sea was something all commanders would avoid like the plague. At Vigo, Spain, maximum pressure from London ensured that the port was closed to the Second Pacific Squadron; the fleet could even be interned. The Spaniards insisted on acting within their rights. And the Russians' fuel situation was critical. The fleet had steamed 1,800 miles and the coal bunkers were low. Finally the port commandant allowed a maximum of four hundred tons per ship to be loaded, a half load. This could signify a cruising range of perhaps no more than 1,500 miles—not much in terms of the total distance the fleet had to travel. There was also a recognition in Whitehall of the superior claims of diplomatic amity with Russia, with the new tripartite European alliance in the offing, and tempers cooled. Rozhestvensky set off for the west coast of Africa. Thence he would take his fleet around the Cape of Good Hope and into the Indian Ocean, not risking the shorter passage through the British-controlled Suez Canal.

Nevertheless, more than problems of morale, or gunnery standards, or the depreciating condition of his ships, the problem of coaling was what would plague Rozhestvensky right to the very end. Between the time when steam took over from sail and when Winston Churchill, as Britain's first lord of the Admiralty, switched the Royal Navy over to recently acquired Middle East oil, all ships were fueled by coal—whole mountains of it, filthy and inefficient stuff. And not all coal was suitable. For instance, that which was mined in Japan was of too poor quality, so just prior to the outbreak of war, the Japanese navy had providently stocked up with 650,000 tons of the best Welsh coal, topped up obligingly by Britain with another 650,000 tons as the war began. But even so, with the fleet consuming 10,000 tons a week at even modest speeds, Togo had to watch his stocks carefully as the war dragged on into its second year, and as he waited anxiously for the arrival of Rozhestvensky's fleet.

The situation was even more worrying for the Russians, never quite sure when, or how, their next delivery would arrive. Cruising at twelve knots, a modern battleship like the French-built *Tsesarevich* with its powerful Belleville engines would eat up three tons an hour; increase the speed to eighteen knots, as under battle conditions, and consumption would soar to fifteen tons. When, during the Battle of the Yellow Sea in 1904, the ship had its funnels riddled with shell fragments, consumption rose to twenty tons per hour; which was why, down to its last tons of fuel, *Tsesarevich* had been forced to seek refuge, and internment, in the nearby neutral port of Kiaochow. Similar rises in consumption would occur when ships suffered from hull-fouling and excess weight, both of which contingencies would invariably affect Rozhestvensky's ships on their long cruise through tropical waters, as we shall see.

After leaving Vigo, the Russian fleet enjoyed a few days of relative calm. At Gibraltar Rozhestvensky ordered Admiral Dimitry von Felkerzam to take his division of older, slower ships through the Suez

Canal, rather than risk the voyage around the Cape of Good Hope. They were to rejoin the main fleet at Madagascar. There was some controversy in the ward room as to the reasoning behind Rozhestvensky's decision. At Tangier the Russian fleet was treated royally by the sultan of Morocco, who was uninterested in the extant state of war. There the sailors were overjoyed to welcome the delivery of a thousand tons of frozen meat. Less pleasing was the coaling from the waiting German colliers that took place in a subtropical rainstorm. Rivers of water poured down the mountains of good Welsh coal, forming filthy streams on the decks of the warships. A dozen ships' bands blasted away to encourage the backbreaking manual work of shifting the heavy bags of coal, "white teeth gleaming through black-stained faces." Rozhestvensky offered a prize of 1,500 rubles for the crew loading the most coal in the shortest time.

The next stop for the Second Pacific Squadron was French-owned Dakar, on the bulge of Africa. With temperatures of up to 120°F (49°C) in October, and with hideous humidity of over 90 percent, the environment for coaling became far more unpleasant. There were ten colliers waiting, bearing thirty thousand tons, and there could be no delay because of the uncertainty of French hospitality. Rozhestvensky, with secret information that the French had forbidden the use of the next port down the line, Libreville (in present-day Gabon), ordered that those of his battleships with a bunker capacity of 1,100 tons would each take on no less than 2,200 tons. His captains were dismayed; how could they conceivably keep their ships clean? Rozhestvensky refused to budge. Draconian instructions were issued: the excess coal should be stored "in any spare place on the upper deck, lower deck, gun deck, poop and in the cockpit, over closed watertight manhole covers, in the bath rooms, drying rooms . . . loose on the quarter deck, with some means to prevent it falling overboard."

Even officers' cabins, up to the rank of commander, were not to be excluded. It would be hard to imagine the natty Captain Pakenham or an Admiral Beresford accepting such conditions. But Rozhestvensky was ruthless, and cleanliness was about the least of his worries. In the gangways, mess decks, and cabins there was an atmosphere similar to that of a mine shaft in a heat wave. For the next four months the Russian crews worked and ate with the bitter fumes in their nostrils, and the filthy and irritant coal dust crept even into their food. Under extreme pressure from the French colonial authorities, coaling at Dakar was accelerated so that it could be completed in twenty-nine hours. Men died of heat exhaustion, including the son of a Russian ambassador, young Lieutenant Nelidov. There were the first signs of mutiny, aboard the supply ship *Kamchatka*, the rogue vessel responsible for the Dogger Bank incident. Like Captain Bligh of the HMS *Bounty*, Rozhestvensky threatened to cast the lot off in open boats. Meanwhile their excessive gross tonnage left the warships verging on the unmanageable, "like sluggish, overloaded river barges" in the words of one chronicler.

But there was worse to come, at Whale Fish Bay (in present-day Namibia) on December 11. There, the Portuguese, Britain's "oldest allies," were unfriendly. Then, because of the inadequacy of the harbor, the fleet had to carry out coaling in open waters, and in a heavy sea. Rolling and pitching, the plucky colliers repeatedly bashed against the sides of the warships as they struggled to come together. The barrels of *Suvorov*'s light guns holed the sides of her own collier, damaging her own guns and torpedo booms. Eventually the attempt had to be called off, and instead Rozhestvensky ordered that the coal be shifted by means of ships' launches. It proved a painfully laborious business, and even after working night and day, only a few hundred tons had been loaded. With his notably short fuse, Rozhestvensky flew into a terrible rage, adding to the barely tolerable burdens of his staff. A young sublieutenant on the *Oryol* lost his mind and ran round the deck, sobbing, "The Japs are waiting for us. We shall all be sunk! We shall all be sunk!" He was locked in his cabin under guard. His ravings and their dark forebodings could be heard throughout the ship, having a bad effect on the crew, who had now been at sea for two months without any news from home.

Then, as the storm died down and coaling could proceed, on December 15 Rozhestvensky received from an itinerant German officer the worst possible tidings from the Far East: Port Arthur's key height, 203 Meter Hill, had been taken. It meant nothing to Rozhestvensky; the German explained that it dominated the whole harbor and could well mean the end of Port Arthur. There was further bad news in the shape of reports that the Japanese had fitted out civil schooners with torpedo tubes, and that these camouflaged vessels might be lying in wait off Durban on the east coast of South Africa. Unrepentant about his Dogger Bank interlude, Rozhestvensky in a fury declared that he would "ruthlessly destroy" all Durban fishing craft that might come within torpedo range. Thus completed what was held to be the worst coaling episode of the entire odyssey.

Amid further stormy seas the fleet ploughed on, around the Cape of Good Hope. It was almost halfway through its long saga. Nerves continued to be stretched, though the rumored schooner attack never materialized. But *Kamchatka*, the albatross of the fleet, its worst ship, and cause of the Dogger Bank melee, once again fell foul of Rozhestvensky, signaling that she had shipped bad coal and wanted permission to throw overboard 150 tons. Acidly Rozhestvensky signaled back: "Find the guilty ones and throw them overboard instead!" A short while later *Kamchatka* followed up: "Do you see the torpedo boats?" Action stations were called, officers roused from their sleep. Then *Kamchatka* followed up: "Sorry." The signalman had used the wrong signal, intending simply to say, "We are all right now." Rozhestvensky's language would have been blistering.

Without further incident the fleet reached Madagascar, anchoring off Île Sainte-Marie, an insalubrious little island off its eastern seaboard used by the French as an overflow penal settlement for Devil's Island.

It was here that Rozhestvensky was to be joined by Admiral Felkerzam's detachment after it had navigated the Suez Canal. But Felkerzam turned up at Nossi-Bé (Nosy Be), some five hundred miles on the other, northwest side of the island, and announced that, on account of their "tired machinery," his ships would be unable to sail for another two weeks. Rozhestvensky flew into another of his famous rages, shouting, "If they are so old that they can't steam then they may go to the devil. We have no use for rubbish here. . . . However, I'll go there myself—I'll dig them out fast enough!" He set off to confront Felkerzam himself. Meanwhile, an oceangoing tug brought calamitous news. Following the fall of 203 Meter Hill, the whole of the First Pacific Squadron had been destroyed inside Port Arthur's basin, and Port Arthur itself was on the verge of capture. All of a sudden it seemed as if the whole purpose of Rozhestvensky's epic voyage had been rendered null and void: his squadron could no longer reinforce the First Pacific Squadron if it had ceased to exist; it could not save Port Arthur if it had already surrendered.

At the same time news came that, at home, a new fleet with the grotesquely ambitious title of the Third Pacific Squadron was being formed—out of all the slow, old ships that Rozhestvensky had expressly left at home. In command would be Rear Admiral Nikolai Nebogatov. Rozhestvensky was to await their arrival in two to three weeks' time. It was the last straw. He dispatched a signal to Saint Petersburg, asking to be relieved of his command; then, feeling suddenly old and ill, he retired to his cabin and locked the door. Needless to say, his request would be turned down by the tsar.

A miserable Christmas was passed in wretched conditions off the Madagascan isle of Nossi-Bé (now its leading holiday resort), with its sweaty, well-named port of Hell-ville. Deep gloom settled in on the personnel of the battleship *Suvorov*, as recorded by Commander Vladimir Semenov. There was now a harsh recognition of the realities of what lay ahead. What had appeared "to be hopeless enough, was suddenly seen as suicidal. Nothing could save them now." Fatalism gripped even the most loyal, such as Semenov, who wrote in his diary, "Saint Petersburg had grasped how utterly hopeless—not to say criminal—our adventure was, and if they had sent us categorical orders to come back, I should have said 'The Lord be praised.'" The arrival of Felkerzam's ancient ships, which had enjoyed an easier cruise through the Suez Canal, did nothing to restore morale; if anything, lacking discipline, they spread more gloom in the fleet.

Then, on Christmas morning, Rozhestvensky stood atop the aft twelve-inch turret and, champagne glass in hand, "as if forcing himself out of a coma," delivered a clarion call to the assembled sailors. He exhorted them, "what fine fellows you are," to do their duty—to Russia. In a voice broken with sobs, he added, "May God help us to serve her honorably, to justify her confidence, not to deceive her hopes. To you, whom I trust! To Russia!" Such an appeal to patriotism, to love of Mother Russia, was sure to provide a shot in the arm. As Rozhestvensky sat down, there were loud cheers, caps thrown in the air and cries of "We'll do it!" "We won't give in!" "Lead us! Lead us!" The Russian fighting man, exhausted and exploited as he might be, was still a formidable foe. But the new year brought further bad tidings: definitive news of the fall of Port Arthur.

Once again the agony of coaling began—the same killing heat, the same plague of coal dust pervading everything. Two seamen on the battleship *Borodino* died of the poisonous fumes. Sunstroke killed another sailor, and knocked out dozens more. Dysentery broke out on several ships. On the *Ural* one officer was killed and another injured by a swinging crane. Ships broke down at regular intervals, causing the fleet to ease its speed on occasions to six knots. Then followed the more welcome activity of revictualling: fifty thousand cases of potatoes from Cape Town; a thousand bullocks, flour, vegetables, chocolates, and biscuits; and fancy foodstuffs and drink for the officers' messes. It would have to last for most of the rest of the journey northeastward.

The state of Rozhestvensky's fleet was hardly improving. Ships stank from rotten food. The cold-storage supply ship *Esperance* had her refrigeration plant break down, and in consequence was forced to throw overboard seven hundred tons of bad meat. The windfall lured in swarms of sharks from miles around; they made life impossibly dangerous for the underwater crews sent down to attempt to clean the ships' keels fouled by weeds and barnacles. Several lives were lost, and this essential work was abandoned. A supply ship, the *Irtysh,* arrived from Russia, supposedly bringing ammunition for the heavy guns; in fact she contained quantities of fur-lined winter coats and twelve thousand pairs of heavy boots—a cargo that was hardly welcomed in the tropical heat of Madagascar. A vital firing exercise, badly needed by the untrained gunners, was a disaster: ships nearly collided, and the gunnery was wild. One of the battleships' munition hoists jammed; inside it a cobra's nest was discovered, the snake insouciantly coiled within. It appeared to have come aboard in a bale of hay for the livestock. To preserve valuable ammunition, which *Irtysh* had failed to bring, Rozhestvensky suspended the exercise, remarking savagely: "By day the entire squadron did not score *one single hit* on the targets which represented the torpedo boats, although these targets differed from the Japanese boats to our advantage, inasmuch as they were stationary."

"Unpardonably bad!" and "Wretched performance!" were among the printable epithets that the fierce commander in chief would employ repeatedly—and with every reason. On one occasion a gun crew was reduced to tears when a poorly machined shell jammed in the breech. On a firing exercise with precious torpedoes, out of seven fired, one jammed, four darted off eccentrically at right angles either to port or starboard, and one went around and around in circles, "popping up and down like a porpoise," and causing consternation throughout the fleet. That would be the last fleet exercise with live ammunition.

In the steamy idleness and boredom of Nossi-Bé, morale continued to plummet. Mutiny broke out on several ships because of the rotten food. Several leaders were hanged from yardarms. Men broke down and cried when they received infrequent letters from home. Drink took over, with lethal brews of kvass made from rye flour and malt. Men fell overboard from drunken orgies. "Our men despaired of escaping from the war with their lives," recorded the seaman Alexey Novikov-Priboy of the *Oryol*, "so they drowned thought in drink, dicing, and drabbing." To Hell-ville women flocked as if by magic from all over the island. Whorehouses and bars suddenly proliferated. The mixed-race owner of one such, called the Parisian Café, announced that he was retiring to Paris "after you all leave." But his joy was cut short.

Abruptly on the morning of March 17, 1905, after two months' enforced idleness waiting for Nebogatov and his collection of rust-buckets—*samotopy* in Russian, or "self-sinkers"—the Russian fleet lifted anchor and departed. Rozhestvensky had received a communication full of menace: the Japanese were threatening that "any colliers discovered in the vicinity of the Russian squadron would instantly be fired upon and destroyed." Reacting promptly, the Hamburg colliers warned that they had signed on to sell coal, not to engage in naval warfare. This prompted Rozhestvensky to leave unhealthy Nossi-Bé forthwith and make a secret coaling rendezvous across the Indian Ocean. Four more colliers would meet him in distant Saigon, carrying a further thirty thousand tons. That should have sufficed to take the fleet to Vladivostok. But weeds and barnacles had prospered in the tropical heat, fouling hulls, slowing speeds by at least two knots, and at the same time increasing coal consumption at cruising speed by between 100 to 130 tons daily. That could prove critical.

Meanwhile, details had dribbled through to the fleet of the appalling events in Saint Petersburg on Bloody Sunday, which would be but an overture to the bloodiest year ever experienced under Romanov rule to date. There were crew members who feared that wives and even children had been among the casualties. Mutiny broke out first on the armored cruiser *Nakhimov.* Nominally it had been sparked by a lack of bread, and by moldy biscuits, but the sailors' grievances were more than that. Mutineers prepared to rush the bridge, then saw that *Suvorov*'s guns were pointed at them. Fourteen ringleaders, chosen arbitrarily, were shot. Rozhestvensky was ruthless in his stamping out of the revolt. Courts-martial were held on board *Borodino*, *Alexander III*, and the regularly miscreant *Kamchatka* (she had excelled herself once more in firing in error a live shell, which bounced off her neighbor, *Aurora*). The worst offenders were shipped home in miserable conditions aboard an ancient vessel Rozhestvensky wanted to be rid of, the *Malay*.

The biggest worry to loyal officers like Commander Semenov was the health of the commander in chief. Semenov noted how increasingly Rozhestvensky would retire to his cabin for hours on end after one of his fits of rage. Sometimes his leg seemed to drag; could he possibly have had a ministroke? And what would happen to the fleet if he collapsed or died? It was an unimaginable prospect. His successor by seniority would have been Rear Admiral Nikolai Nebogatov, currently plodding along with the Third Pacific Squadron of rust buckets, to join up shortly with Rozhestvensky's fleet and, in theory, replace the First Squadron, lost at Port Arthur. There could hardly have been a greater contrast between the two men. At fifty-six, Nebogatov was fat and small, with a chubby face suffering from chronic eczema, though partly concealed behind a short beard. Moving about his deck in little steps, he was an unimpressive figure, but quietly spoken, mild-mannered, approachable, and tolerant of his crews to an extent incomprehensible to the fiery Rozhestvensky, to whom he was anathema. At Port Said he had telegraphed Saint Petersburg for instructions. Back came a vague response: "You are to join up with Rozhestvensky, whose route is unknown to us." Having crossed the Indian Ocean safely to reach Singapore, he received news there that Rozhestvensky had already passed through, on April 8.

From when it had left Nossi-Bé until the evening of April 8, Rozhestvensky's squadron had in fact been lost to the world, as well as to the Russian Admiralty. Nearly fifty ships strong, it had steamed across the wide, lonely Indian Ocean for 3,500 miles without seeing another ship. It had only had faint sight of land at Sumatra, where it chose the narrow Straits of Malacca to attain Singapore. To Semenov's expert eyes, their safe passage was "unexampled in the history of steam navigation." Overcome with depression and anxiety, several more sailors had thrown themselves overboard. Every four or five days there would be a halt for the backbreaking task of coaling from Rozhestvensky's own emergency colliers, and more frequent stops on account of breakdowns. The food situation was dire; long ago the fleet had run out of cabbage for making the soup that was essential to a Russian sailor (though hardly palatable to his Royal Navy counterpart). Ships' cooks had had to make do with manioc, a poor substitute. There were more cases of rotting meat and worm-ridden biscuits, and more angry murmurings.

When it arrived at Singapore on the evening of April 8, literally out of the blue, the sight of the great fleet caused the utmost excitement. Even *The Times*, somewhat amending the hostile tone it had adopted after the Dogger Bank incident, rated it "a splendid spectacle," as impressive as anything the British navy had ever presented. "The smoke they made was visible for miles," reported Reuters. "The ships, magnificent but foul, were proceeding at about eight knots, and it took them fifty-five minutes to pass a given point. All the vessels showed signs of their long voyage in tropical seas, about a foot of seaweed being visible along the waterline, and the decks were laden with coal." Another London leader-writer declared that an account of the fleet's passage "will send a thrill of admiration through all Englishmen who read it." Would Togo, one wonders, on reading such reports, have been affrighted—or encouraged?

But Rozhestvensky, regarding British Singapore as essentially a hostile port, whence Togo would certainly receive detailed intelligence, decided not to tarry there. He stayed just long enough to receive some more unwelcome news, shouted at him through a megaphone by the Russian consul aboard a steam launch: "Mukden has fallen . . . General Kuropatkin has been dismissed." Next came information of much greater immediacy: "Admiral Kamimura's cruiser squadron called at Singapore three days ago, and is now believed to be on its way to North Borneo, and twenty-two more warships under Togo's flag . . . are now at Labuan* . . ."

Other news, of personal concern to Rozhestvensky, and indeed causing him great affront, was that, by an order from Saint Petersburg, he was to hand over command on reaching Vladivostok to an Admiral Aleksei Birilev, who was already trundling eastward along the Trans-Siberian Railway. To have taken his great fleet safely halfway around the world, possibly yet to face a major battle with the enemy, then to be rewarded with the sack was hardly the way to treat a commander on the eve of his greatest challenge. Understandably, Rozhestvensky sank into the deepest Russian gloom. There were also rumors that the United States was putting pressure on the belligerents to end the fighting. The forceful president Theodore Roosevelt was hatching a peace plan; Japan was considering its terms. So what, Admiral Rozhestvensky and his brave Russians may have wondered, was now the point of sailing on to face a murderous battle when peace was already in the offing?

As they headed for Cam Ranh Bay through the South China Sea, the squadron now passed many ships, including more British naval vessels, which they thought must be tailing them on behalf of their Japanese allies. Why, Commmander Semenov frequently asked himself, were the Japanese missing so many opportunities for a surprise

*An island off the northwest coast of Borneo, Labuan was menacingly close to Rozhestvensky's onward route.

attack? What was Togo doing? The answer was, once repairs to his damaged vessels had been completed, the Japanese leader—more or less at leisure—was training on, perfecting the maneuvers of his fleet, carrying out gunnery exercises such as Rozhestvensky had so noticeably failed to achieve in the Indian Ocean. Cannily, to preserve his heavy ammunition and gun tubes, he had carried out firing by the use of subcaliber adaptors, which allowed the use of lighter shells. Yet, like the Russians, Togo too had to ration exercises; his coal supplies were also finite and waning. But he had the unassailable advantage of time, fresh manpower, and morale on his side. As Napoleon had once remarked, "The moral is to the physical as of three to one."

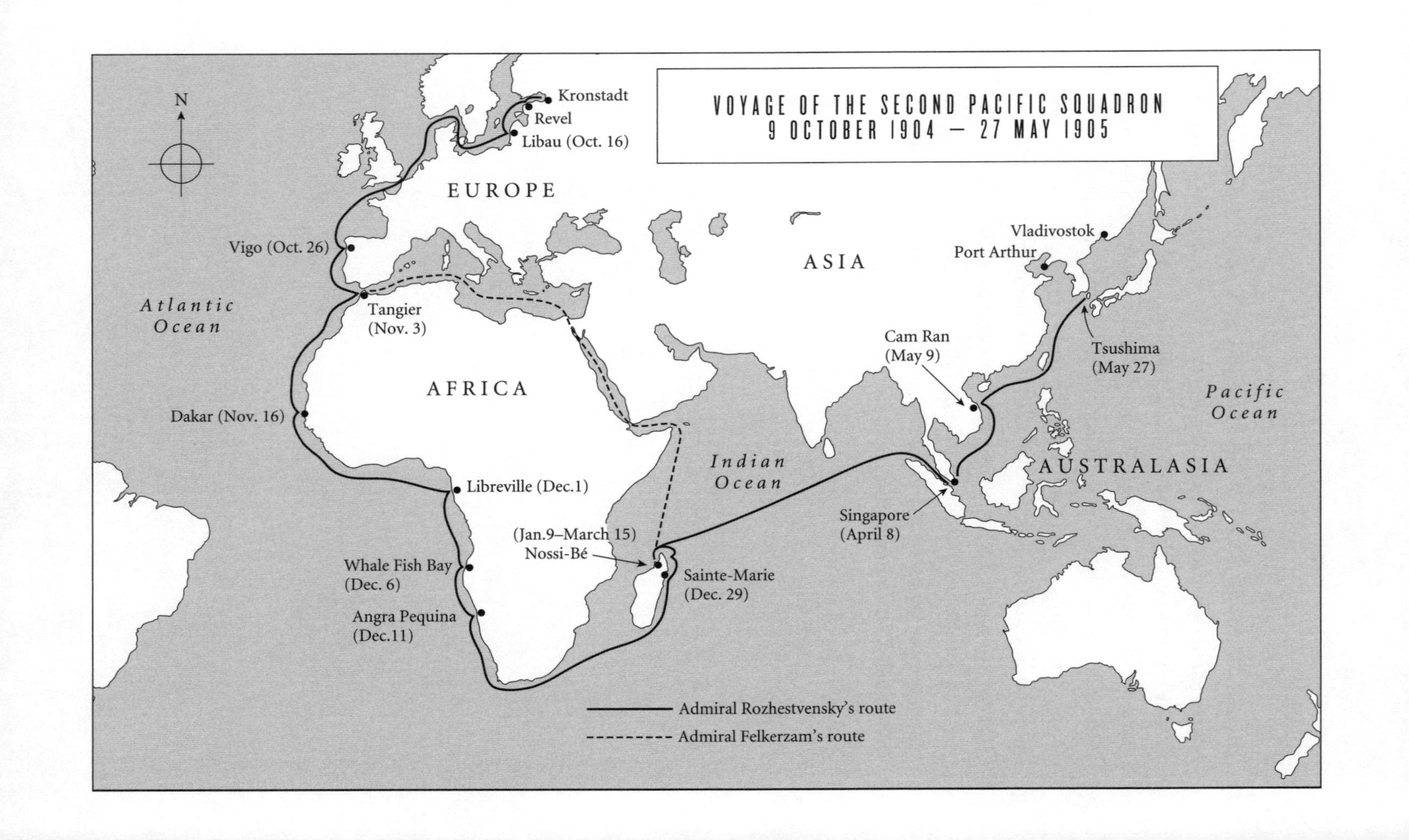
VOYAGE OF THE SECOND PACIFIC SQUADRON
9 OCTOBER 1904 — 27 MAY 1905
N
Kronstadt
Revel
Libau (Oct. 16)
EUROPE
Vigo (Oct. 26)
Atlantic Ocean
Tangier (Nov. 3)
AFRICA
Dakar (Nov. 16)
Libreville (Dec.1)
Whale Fish Bay (Dec. 6)
Angra Pequina (Dec.11)
(Jan.9–March 15) Nossi-Bé
Sainte-Marie (Dec. 29)
Indian Ocean
ASIA
Port Arthur
Vladivostok
Cam Ran (May 9)
Tsushima (May 27)
Singapore (April 8)
Pacific Ocean
AUSTRALASIA
Admiral Rozhestvensky's route
Admiral Felkerzam's route

CHAPTER 5

The Battle at Sea

ON HIS LAST coaling and rest-up rendezvous before the final confrontation, Rozhestvensky assembled his fleet in the great natural harbor of Cam Ranh Bay in Indo-China. It was the first week of May 1905. In what is now Vietnam, during the war of that name Cam Ranh was host to many a US fleet, but in 1905 it still flew the flag of the French empire. Once again Rozhestvensky and his huge armada had disappeared from sight of reporters and other navies. Yet in April a resourceful editor of the London *Daily Telegraph*—in the best tradition of Evelyn Waugh's Lord Copper of the *Beast*—had wired his new man in Hong Kong, William Donald, a go-getting Australian with an instinct for a story: RUSSIAN FLEET REPORTED LOST SINCE LEAVING RED SEA STOP MAY BE SOMEWHERE IN YOUR AREA STOP GO FIND IT.

Accordingly, acting on a hunch, Donald booked a steamer for Saigon. Passing Cam Ranh, he spotted the telltale forest of smoke columns; pressing forward, he found himself confronted by the massive hulk of the Russian flagship *Suvorov*. On the flank, as if guarding the Russian fleet, lay a French cruiser, *Descartes*. It was May 1; Rozhestvensky had arrived on April 14. The *Telegraph* had a world-beating scoop, and Japan's loyal ally, Britain, a piece of prime intelligence. As Russia was in a state of war, Rozhestvensky's fleet was not allowed to remain in any port for more than twenty-four hours. Rozhestvensky was in the middle of his crucial coaling operation, but was ordered to leave forthwith by a reluctant France. This he did, but after a brief sortie he immediately returned. The pause at least enabled him to link up, finally, on May 14 with Nebogatov and his fleet of tired old "flat-irons and galoshes." There was a solemn though brief moment as the two squadrons, now over fifty vessels strong and sixteen thousand miles away from home, steamed majestically toward each other. Briefly the leaders celebrated Russian May Day with a crate of Mumm, toasting the health of the tsar "and victory." It was, however, the coolest of encounters, this meeting of the two Russian admirals, who disliked and even despised each other. Nevertheless, news of this great event, reaching Saint Petersburg, gave the jingoists one last fleeting hope of a victory in the offing; despite all the setbacks, invincible Mother Russia was about to regain her rightful claim to be regarded as "mistress of the Pacific."

In Cam Ranh, however, the triumphant juncture of the two Russian fleets was seriously diminished when the obese Admiral Felkerzam suffered a stroke. Rozhestvensky managed to keep this quiet. The significance of this mishap was that, if Felkerzam were eliminated, the despised Nebogatov as next in line would automatically become second-in-command of the whole enterprise. Rozhestvensky displayed his contempt for Nebogatov by receiving him with icy formality at Cam Ranh, refusing to discuss his plans: "He gave me neither instructions nor advice. . . . I only saw Rozhestvensky once in the whole course of the cruise." This was hardly a happy omen for the decisive engagement that lie ahead.

Meanwhile, the eagle-eyed Donald was not fooled: "These ships," he reported, "will be disgraced in their first engagement." A second, follow-up story in the *Daily Telegraph* reported Rozhestvensky's return to Cam Ranh and the news that he had then been ordered by the French, definitively, to pack up and leave French waters.*

---

*So impressive was Donald's reporting that the *Daily Telegraph* felt able to claim his forthcoming account of the Battle of Tsushima as an "eyewitness's story," although it would in fact be written from the tranquil setting of Hong Kong, 1,200 miles to the south.

The interruption of Rozhestvensky's vital, final refueling meant that he would now not have enough coal to take the longer and safer route to Vladivostok by sailing to the east of Japan, rather than through the perilous Scylla and Charybdis of the Tsushima Straits, which were narrow and commanded by the Japanese home fleet. A decisive factor was that, through what seems like a foolish housekeeping error—probably a result of campaign fatigue from the long, long journey—one of Rozhestvensky's most powerful units, the battleship *Alexander III,* which had consistently won the "fastest coaling" prize throughout the previous weeks, loaded only three hundred tons of coal at Cam Ranh, instead of nine hundred. There was no way that would suffice to take the *Alexander III* on the longer, safer route to Vladivostok. It was a mistake that seemed to indicate the state of morale within Rozhestvensky's armada as a whole. Once again the awful, choking coal dust was everywhere. Day after day the number of suicides multiplied. Aboard *Oryol* drunk and mutinous crewmen broke open a case of liquor and attacked officers. Morale was hardly helped by the news trickling through of the mounting discontent at home, and of the smashing defeat of Kuropatkin's army at Mukden. ("We shall have about as much chance as a gamecock would have in a battle with a vulture," declared an engineer officer aboard *Oryol* on hearing the dire tidings from Port Arthur.) What, indeed, was the point of driving on into the jaws of death now? The remarkable thing, surely, was that there remained a single ounce of fighting spirit among Rozhestvensky's Pacific squadrons.

Aboard his flagship *Mikasa*, Togo was able to deduce Rozhestvensky's course and timing from Donald's two dispatches, with reasonable certainty. In tranquil Japanese waters, he waited anxiously—but with confidence. On May 14 the great combined fleet was reported to have left Cam Ranh Bay, heading north. With no radar, no spotter planes, and radio over only very limited ranges, Togo had to depend on visual sightings and good intelligence analysis.* Once again the elusive Rozhestvensky and his armada seemed to be swallowed up in the vastness of the ocean. Only on May 25 did a report of six Russian auxiliaries, including their colliers, arriving at Shanghai confirm to Togo that the enemy was heading for the Straits of Tsushima. Had it been either of the longer routes to Vladivostok, the La Pérouse Straits between Hokkaido and Sakhalin Islands, or the Tsugaru between Hokkaido and Honshu, Rozhestvensky would have been sure to have kept the colliers with him as he sailed eastward around Japan.

So Togo kept his fleet in waiting, undivided and concentrated, at the port of Masan, on the southernmost tip of the Korean peninsula. It was a secret base, in a deep and sheltered bay, that Togo had been developing for many months. Apart from a few light forces held back to defend the home ports, the whole of Japan's navy was concentrated there. The cautious admiral delivered to his ships an uncharacteristically aggressive message on May 15: "If your sword is too short take one step forward. . . . There is no need to think of defence. A positive attack is the best form of defence . . ." Admiral Hikonojo Kamimura was dispatched to watch, and heavily mine, the approaches to Vladivostok, should any of Rozhestvensky's ships break through the Tsushima cordon. Only the battleship *Asahi*, with an impatient Captain Pakenham on board, champing for action, had been temporarily left in the Inland Sea, the damage inflicted at the Battle of the Yellow

*In 1905, ship-to-ship radio communication was still in its infancy. The Russians had equipped their latest, *Poltava*-class battleships with the most up-to-date German-designed radio sets. But each weighed up to six tons and, with a fraction of the power of today's mobile telephone, could communicate—intermittently—over only a maximum of forty miles. Togo had equipped his fleet with the latest Marconi wireless sets, using Morse code, but they had limited reliability and a maximum range—on a good day—of fifty miles. Nevertheless, Togo depended on them for locating and shadowing the Russian fleet as it sailed up toward Tsushima, thus giving the forthcoming battle the distinction of being the first at sea to be fought with electronic warfare.

Sea the previous August still being repaired. But "nothing could be better than the temper of the personnel of the Japanese Navy or the condition of their ships" according to the final pre-battle report from the Royal Navy's observers.

For the Russians, the night of May 26–27 (May 13–14 on the Russians' Julian calendar) brought just the kind of North Pacific weather that they might have prayed for: thick mist and dense patches of fog. Every hour that it enveloped them brought the fleet another hour closer to Vladivostok and safety. In his Fleet Order No. 294, Rozhestvensky prayed "that God may strengthen my right hand, and that if I fail to fulfil the oath I have sworn, He may purge my country from shame with my blood."

He now reorganized his ships into three divisions, the first consisting of the four new *Borodino*-class battleships commanded by himself. Admiral Dimitry von Felkerzam, though mortally sick, commanded the Second Division of the battleships *Oslyabya* (flagship), *Navarin*, *Sissoi Veliky*, and the armored cruiser *Admiral Nakhimov*, while Nebogatov retained his ships as the Third Division. As just one more setback to hit the doomed fleet, the obese Felkerzam died on May 26; but, in order to maintain morale, Rozhestvensky elected not to inform the fleet. So the Second Division nominally went into battle led by a dead admiral, though the captain of *Oslyabya* in fact became its commander, while Nebogatov had no idea that he was now the squadron's de facto second-in-command.

Approaching the Tsushima Straits, whose banks were just thirty miles across, Rozhestvensky ordered the vast column to swing around on a more northeasterly course, to north 23 degrees east, for Vladivostok. It was the home stretch. The Russian fleet was now sailing in six separate columns, which, in the conditions of severely restricted visibility, made command and control difficult. There were last-minute preparations for battle. Anything that was combustible (excepting the emergency stocks of coal)—bedding, hammocks, furniture—was thrown overboard. The guns were sprinkled with holy water, letters written home, and the surgeons prepared for casualties. A welcome tot of vodka was issued to every man. It would not have been very different from last-minute preparations aboard Nelson's *Victory* a hundred years before, the prospects equally terrible for the losing side. That night before the battle, "hardly anyone slept; it was too clear we'd be meeting the enemy in full strength. . . . The orderlies secretly tell the officers that rats have appeared on the accommodation deck," recorded Flag Captain Clapier de Colongue, Rozhestvensky's handsome and rather effeminate chief of staff aboard *Suvorov.*

On the Japanese side, as soon as he realized that the enemy was committed to taking the shorter, Tsushima route, Togo ordered all but the essential minimum of his precious stocks of Welsh coal to be dumped overboard, to ensure the maximum speed and maneuverability of his ships. Together with the excess of seaweed fouling the hulls of the Russian ships after their eighteen-thousand-mile journey, and the weight of their extra coal supplies, this would give Togo at least a knot or more of extra speed over his adversary. On top of this, Rozhestvensky's emergency supplies of coal rendered some of his battlewagons so low in the water that the operation of their hull-mounted secondary armaments were threatened in the event of heavy seas.

Suddenly, at 0245 Japanese time, on May 27, there was a brief clearing in the fog. Through it a Japanese auxiliary cruiser, *Shinano Maru*, observed three lights twinkling on the distant horizon. She closed in to investigate. They were the lights of a Russian hospital ship, *Oryol* (not to be confused with the battleship of the same name). In compliance with the rules of war, the hospital ship continued to display its lights. Mistaking the *Shinano Maru* for a friendly vessel, she failed to notify the Russian command; instead, obligingly, she signaled *Shinano Maru* the useful information that there were other Russian vessels nearby in the fog. *Shinano Maru* then made out the shapes of ten other enemy ships. The cat was out of the bag, the Russian Baltic Fleet discovered. At 0455, Captain Narukawa of the *Shinano Maru* was signaling Togo at his Masampo base: "Enemy is in Square 203."*

On the Russian side, following this first contact, all was peace and tranquility—for a few hours. No more Japanese ships were seen. Optimists hoped that, having thrown them off the scent in the fog, the Russian fleet had escaped. Intercepting the Japanese wireless traffic, Rozhestvensky knew better. At 0505 Togo was ordering his whole fleet, of over forty vessels, into the Straits of Tsushima. He was in the lead, aboard his flagship *Mikasa*, the ship's band loudly playing the Japanese national anthem on the aft deck. At 0634, the navy minister in Tokyo received the following signal: "The enemy fleet has been sighted. Our fleet will proceed forthwith to sea to attack the enemy and destroy him. Today's weather is fine but waves are high."

About seven hours later, at 1:40 p.m., the main fleets sighted each other and prepared for battle. At 1:55 p.m., Togo ran up his "Z-flag," signaling a simple, Nelsonian message: "The Empire's fate depends on the result of this battle, let every man do his utmost duty." The two great fleets were set on parallel courses, heading toward the northeast. Togo, with his advantage of speed, was clearly aiming to overtake Rozhestvensky. For no very clear reason, Rozhestvensky decided to concentrate his First and Second Divisions into a single battle line, but poor training and confused signaling led to a chaotic muddle. This was still being sorted out when the Japanese battleships appeared from the mists, a long line of gray hulls streaming out black smoke. In contrast, the conspicuous bright yellow funnels of Rozhestvensky's battleships may well have helped the Japanese range finders.

---

*Well in advance of the battle, Togo had gridded the Tsushima area into squares, to help his gunners obtain precise ranges of enemy vessels and lock in on them. By chance, "203" was also the number of the key strategic height dominating Port Arthur, 203 Meter Hill, whose capture had defined the fall of the base to Japan.

Cutting across the Russian line, with the clear intention of crossing the T in the best tradition of Nelson at Trafalgar, Togo ordered his ships to "turn in sequence" at 1405, doubling back across the Russian line so as to strike on its weaker flank. It was a breathtakingly bold move, one that could have been performed only by highly competent seamen. It meant that for a period of roughly ten minutes each Japanese ship would be turning, pivoting on precisely the same spot, making them easy, sitting targets for enemy gunners. But when it was completed, the maneuver left the Japanese battleships in a powerful arc able to bring the full weight of their broadsides to bear on the Russian ships, which could deploy only their forward-pointing main armaments. The risk Togo was taking was immense, and had Rozhestvensky been a Makarov it may well have been disastrous. As it was, it was his own ship, *Mikasa*, which, in the van, bore the brunt of the opening exchanges of fire. Clearly Togo intended to purge any suggestion of the excessive caution he had shown in earlier engagements, such as the first surprise attack of February 1904, and the Battle of the Yellow Sea that same year.

Initially the Russian shooting was more accurate than anticipated at the unusually long range of nine thousand yards. The first shots fell only twenty yards astern of *Mikasa*; thereafter it was hit fifteen times in the next five minutes with both six-inch and twelve-inch heavy shells. His staff urged Togo, who had already been wounded by a splinter in the thigh, to leave the exposed bridge for the safety of the armored conning tower, but he refused, saying, "I'm getting on for sixty . . . and this old body of mine is no longer worth caring for."

A sense of Drake-like coolness was manifest. Aboard *Asama*, Captain Yatsushiro amazed his staff by playing the flute until it was time for the batteries to open fire. On *Asahi*, which had now just rejoined the fleet after being repaired, "officers gathered round

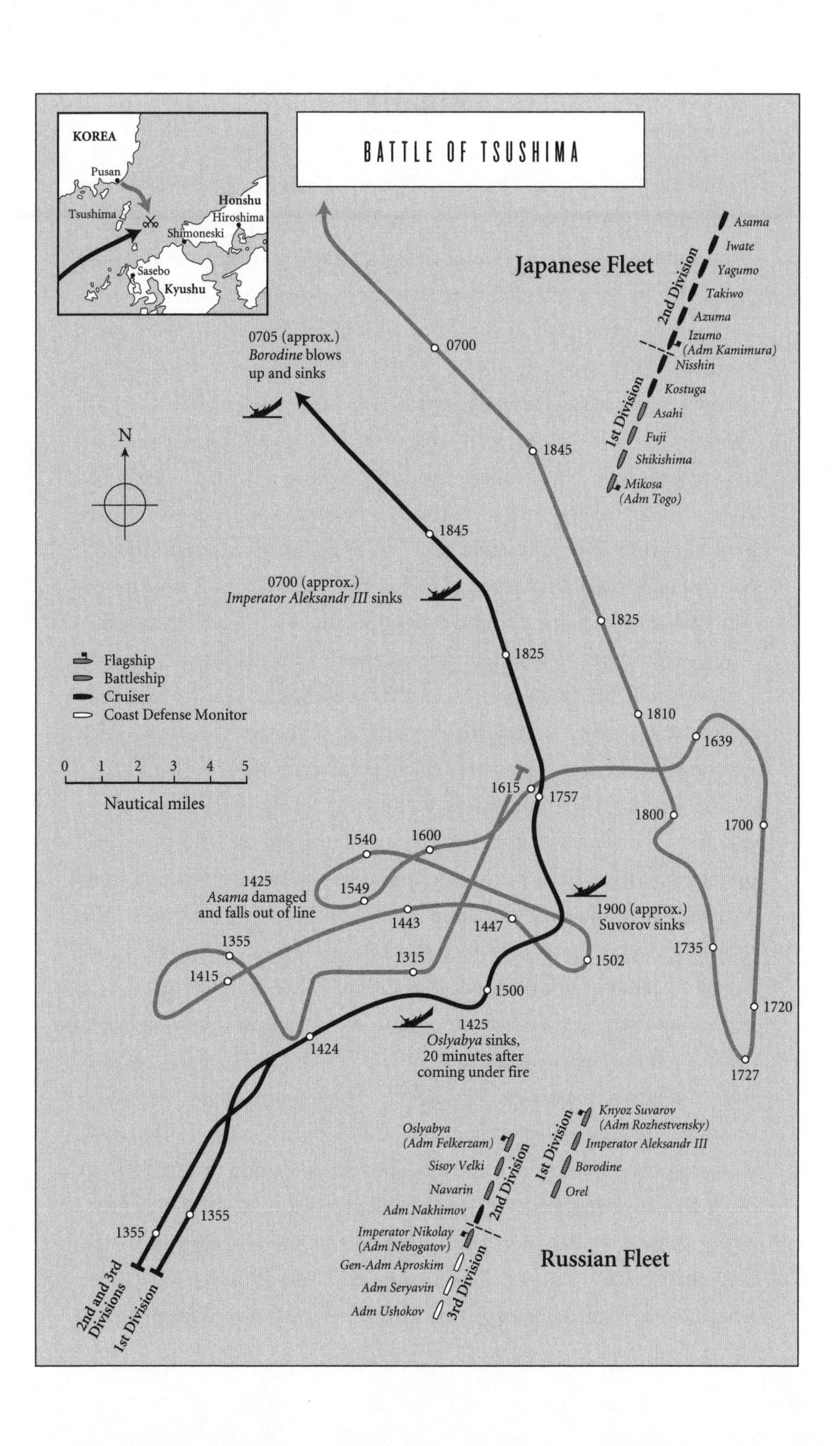

BATTLE OF TSUSHIMA
KOREA
Pusan
Tsushima
Honshu
Hiroshima
Shimoneski
Sasebo
Kyushu
Japanese Fleet
2nd Division
Asama
Iwate
Yagumo
Takiwo
Azuma
Izumo
(Adm Kamimura)
1st Division
Nisshin
Kostuga
Asahi
Fuji
Shikishima
Mikosa
(Adm Togo)
0705 (approx.)
Borodine blows up and sinks
0700
1845
1845
0700 (approx.)
Imperator Aleksandr III sinks
1825
1825
N
Flagship
Battleship
Cruiser
Coast Defense Monitor
0 1 2 3 4 5
Nautical miles
1810
1639
1615
1757
1800
1700
1540
1600
1425
Asama damaged and falls out of line
1549
1443
1447
1900 (approx.)
Suvorov sinks
1355
1315
1502
1735
1415
1500
1720
1424
1425
Oslyabya sinks, 20 minutes after coming under fire
1727
Knyoz Suvarov
(Adm Rozhestvensky)
Imperator Aleksandr III
Borodine
Orel
1st Division
Oslyabya
(Adm Felkerzam)
Sisoy Velki
Navarin
Adm Nakhimov
2nd Division
Imperator Nikolay
(Adm Nebogatov)
Gen-Adm Aproskim
Adm Seryavin
Adm Ushokov
3rd Division
Russian Fleet
1355
1355
2nd and 3rd Divisions
1st Division

the tobacco tray near the after turret, where an observer found them quite confident of victory, smoking cigars and listening to a gramophone." Wrote one historian, "a stranger seeing them would not, I think, have guessed that they were people who this very day were going into battle, a battle in which many perhaps were fated to die."

No one aboard *Asahi* seems to have been cooler than Captain Pakenham, Britain's official observer. In the middle of the battle, he ordered a deck chair brought up and sat on the exposed quarterdeck, taking notes "as though he were aboard a committee boat keeping score at a yacht race." Always immaculately dressed, almost a dandy, Pakenham, with his trim beard and bushy moustache, monocle in one eye and telescope glued to the other, typified almost to the point of caricature a certain kind of Royal Navy officer of the Edwardian era. As a teenage midshipman, he had been commended for gallantry for rescuing a coxswain who had fallen overboard. During the battle, *Asahi* suffered nine hits from heavy shells, which killed eight and wounded twenty-three. Pakenham was hit by a flying fragment that he identified as "the right half of a man's lower jaw, with teeth missing." His uniform splashed with Japanese blood, Pakenham went briefly below, only to return to his deck chair post clad in an immaculate fresh white uniform. Like the Japanese commander in chief, Pakenham seemed to be immune to any sense of personal danger. His notes taken on board *Asahi* provide the best account of the Battle of Tsushima as seen from the Japanese point of view, before the fog of war and the smoke of battle confused everything. As well as publicizing the Japanese victory in Royal Navy circles, his reports confirmed the superiority of Togo's training and tactics. His observation of the dominance of big guns led, in part, to the adoption of the all-big-gun battleship in the Royal Navy, beginning with the 1906 *Dreadnought*, which, under the aegis of Admiral John Arbuthnot "Jacky" Fisher was to become the backbone of the new, reformed Royal Navy by 1914.*

Based on Pakenham's reports, the Royal Navy's *Official Report* weighed up the moral fiber of the two opposing fleets as the battle began:

> on the one side were crews of veterans who had already shown magnificent fighting qualities, who had been handling the huge mass of material in their charge for over a year under war conditions, who were under the command of an admiral in whom they reposed the utmost confidence. On the other side were collections of men hastily assembled, who were deficient in gunnery training . . . the Russian crews were half beaten before a shot was fired. . . .

Pakenham recorded the anxious moments when Togo first crossed the Russians' T, as his ships performed their tricky turns in sequence: "It was possible to see down the length of the Russian Lines. In the right column the four biggest battleships loomed enormous, dwarfing all others into insignificance. . . . It was not easy to realise that the battleships of Japan were probably producing at least equal effect on the minds of the Russians." As the range closed to about seven thousand yards, part of the Russian fire continued to rain down on Togo's *Mikasa*, six-inch shells as well as the monster twelve-inch projectiles, but the main weight of shellfire continued to converge "on the Japanese turning point, and it [was] interesting to watch each ship approach

*In 1916, Pakenham, now promoted to rear admiral under the dashing David Beatty, led the Second Battle Cruiser Squadron during the Battle of Jutland, and in 1917 replaced Beatty as commander in chief of the Battle Cruiser Fleet. Retiring as Admiral Sir William Christopher Pakenham, KCB, KCMG, KCVO, a member of the Longford clan, he died unmarried. For his work during the Russo-Japanese War he was decorated by both the British government and the emperor of Japan (Order of the Rising Sun [Second Class]).

and run through this warm spot, a feat all were lucky enough to accomplish without receiving serious injury." The "luck," judged Pakenham, was aided by the "absolute confusion" of Rozhestvensky's battle line. Two Japanese cruisers were hit (surprisingly, by Nebogatov's despised "flatirons") and forced to retire. Aboard *Mikasa* a twelve-inch gun turret was put out of action by an ammunition explosion caused by unpredictable *shimose* charges.

Nevertheless, the tide of battle turned with furious swiftness. Togo's ships, some capable of turns of speed up to five knots faster than Rozhestvensky's weed-clogged and overloaded vessels, were difficult for the Russians and their outdated (1880s) Liuzhol range finders to track. The Barr & Stroud range finders, introduced in 1903, a year before the war, were speedier and substantially more accurate. The Japanese also had the advantage of Krupp armor plate, and Krupp armor-piercing shells for their twelve-inch guns. And, of course, they had Togo. On the other side, as Richard Hough saw it, the Russians were to lose the battle "by the momentary hesitancy of a tired and wasted Admiral who could not make up his mind, and when he changed it, disregarded the consequences."

A terrifying storm of steel and high explosives now poured down on the Russian battleships. Given the confusion and the smoke of battle, not to mention the lack of survivors, much of the blow-by-blow description of the Russian side of the engagement stems from Commander Vladimir Semenov, aboard Rozhestvensky's flagship *Suvorov*. Many individual acts of great heroism, and also cowardice, would go unrecorded. It was *Suvorov* that bore the brunt of the concentrated deluge of shells from Togo's battle line. Such was now the closeness of the two fleets—perhaps less than five thousand yards apart—that even Togo's secondary, six-inch sponson guns could register hits. Soon Semenov described himself as "slipping and sliding in pools of new blood." One of the first shells exploded with deadly effect in the officers' quarters, starting a fire.

Semenov, a veteran of the previous year's Battle of the Yellow Sea, noted with concern a kind of "stupor which seems to come over men, who have never been in action before, when the first shells begin to fall. . . . The men at the fire mains and hoses stood as if mesmerized, gazing at the smoke and flames, not understanding, apparently, what was happening." He galvanized them into action, to get the hoses flowing. Then he himself was hit and stunned. "Shells seemed to be pouring upon us incessantly." The Japanese shells, equipped with instantaneous fuses and filled with their deadly *shimose* explosive, "burst as soon as they touched anything." Handrails and even funnel guy ropes "were quite sufficient to detonate a thoroughly efficient air burst . . . the splinters caused many casualties. Iron ladders were crumpled up into rings, and guns were literally hurled from their mountings." The exploding Japanese shells produced such a high temperature, he claimed, that

> I actually watched a steel plate catch fire from a burst. Of course the steel did not burn, but the paint on it did. Such almost non-combustible materials as hammocks, and rows of boxes, drenched with water flared up in a moment. At times it was impossible to see anything with glasses, owing to everything being so distorted with the quivering, heated air.

Semenov tried to make his way to the conning tower, where Admiral Rozhestvensky and *Suvorov*'s captain were peering out through a chink between the armor and the roof. He realized they couldn't see what was actually happening aboard the ship, the pride of the Russian navy. "What havoc!—burning bridges, smouldering debris on the decks—piles of dead bodies. Signalling and judging-distance stations, gun-directing positions, all were destroyed. And astern of us the *Alexander* and *Borodino* were also enveloped in smoke."

Men were falling "faster and faster": "The officer commanding the fire parties had had both his legs blown off and was carried below. . . . Over and over again the hoses in use were changed for new ones, but these also were soon torn to ribbons, and the supply became exhausted. Without hoses how could we pump water on to the bridges and spar-deck where the flames raged?" One heavy shell had landed squarely on the ship's hospital, already full of wounded men: "Between the wrecked tables, stools, broken bottles and different hospital appliances were some dead bodies, and a mass of something, which, with difficulty, I guessed to be the remains of what had once been men."

The Russian ships nevertheless kept on firing with unflinching courage, that dogged courage of simple Russian servicemen, such as dignifies their history from time immemorial—men knowing that they were doomed, who would yet fight on like heroes. Unsurprisingly, considering the lack of training, their gunners' accuracy was considerably less than Togo's; observers calculated that the Japanese registered 3.2 hits per hundred shells fired, while the Russians had a success rate of no more than 1.5. Moreover, many of their hits were by shells that proved to be duds. It was reckoned that not a single Japanese ship received a hit after the furious first half hour. At about 1420 hours Togo's ships switched to armor-piercing shells, at increasingly close range, with horrendously effective results.

Aboard *Suvorov* the Japanese shells had also ignited the overloaded stocks of coal, causing dense smoke to form. When it cleared, Semenov attempted to make his way through the wreckage to see what had happened to the *Suvorov*'s after turret, "but along the upper deck no communication between bow and stern was possible." It became evident that "owing to our steering-gear being out of order, we had turned nearly 16 points." At this juncture, in the smoke and general confusion of battle, accounts lose their coherence and become disconnected. But with the admiral's badly battered flagship limping out of line, all notions of the brave Russian fleet as a fighting unit began to dissolve. The Japanese account recorded that *Suvorov* was

> so battered that scarcely any one would have taken her for a ship, and yet, even in this pitiful condition, like the flag-ship she was, she never ceased to fire as much as possible with such of her guns as were serviceable. . . . Her upper part was riddled with holes, and she was entirely enveloped in smoke. Her masts had fallen and her funnels came down one after the other. She was unable to steer, and her fires increased in density every moment.

Nevertheless, she kept on firing with a courage that drew admiration from even the enemy seamen. It was at this point that Rozhestvensky himself was hit, several times, but most seriously with a shell splinter in his skull. He was "carried into the turret and seated on a box, but he still had sufficient strength at once to ask why the turret was not firing. . . . None of us knew how badly he was wounded because, to all enquiries when he was hit, he angrily replied that it was only a trifle."

Meanwhile, "the engines had ceased to work. The electric light had given out for want of steam; and no one came up from below." To Pakenham, "her condition seemed infinitely deplorable." Aboard *Suvorov* it was all too clear that the great ship was dying, and the admiral had to be evacuated by torpedo boat. Half a dozen seamen "got hold of some half-burned hammocks and rope from the upper battery, and with these had begun to lash together something in the shape of a raft on which to lower the Admiral into the water." Rozhestvensky was then "almost thrown, on board the torpedo boat, at a moment when she rose on a wave and swung towards us." The admiral was then transferred to the destroyer *Buiny*.

For all intents and purposes, *Suvorov* was abandoned. Yet one brave young midshipman, a German named Werner von Kursell, a humorous and eccentrically wild youth much loved in the officers' mess, insisted on manning a small-caliber gun on the stern, the battleship's last serviceable weapon, to the very end, endeavoring to keep at bay Togo's jackal-like torpedo boats—a midshipman left in charge of the Russian flagship.

The battleship *Oslyabya* had now been hit by heavy shells right on the waterline. "Three shells, one after the other, almost in the same identical spot," Semenov was told by an officer who survived. "Imagine it! All of them in the same place! All on the waterline under the forward turret! Not a hole—but a regular gateway!" Her ten-inch guns were outranged by the twelve-inch weapons of Togo's fleet. The lethal Japanese shells struck her just fifteen minutes into the engagement and immediately caused serious flooding. Forward motion exacerbated the flooding, so, forced to cut speed, she too had to fall out of line, listing dangerously to port. Ill-advisedly, the commander, Captain Ber, gave orders to flood the starboard magazine in an effort to correct the list. But this just added to the forward weight, with mortal consequences for the ship's stability. Suddenly she rolled over to port, until her funnels were touching the sea, and at about 1515 Captain Ber gave the order to abandon ship. *Oslyabya* sank within minutes, her propeller still turning. Captain Ber and 470 members out of her crew of 770 went down with her.

Oslyabya was the first modern steel battleship to be sunk entirely by naval gunfire. It may well be that the speed of her capsizing had something to do with her bulbous "tumblehome" hull design, grossly overloaded as she was with coal stocks.* The sight of her rolling over was deeply demoralizing to those in the rest of the fleet who witnessed it. For Pakenham, it presented a "terrible spectacle." The Russian fleet now lost any semblance of order: its line gone, each battleship was now isolated from her sister ships. To some observers the engagement now seemed like a "battle of shadows," as the Russian ships lost sight

*Following the lessons of Tsushima, navies would discontinue use of tumblehome hulls.

of each other altogether in the heavy seas and smoke. Similarly, an overall account of the battle now loses all cohesion, as the engagement broke down into individual battles, often involving great heroism on the Russian side. Even Pakenham from his strategically placed deck chair found it impossible to follow events.

Togo now regrouped his battle fleet and moved in for the kill, closing at a range of less than five thousand yards. There was a two-hour pause in the battle while the Japanese admiral reordered his battle line. Then the Japanese ships concentrated their fire on *Borodino* and *Alexander III*, both completed (like their sister ship *Suvorov*) between August 1903 and September 1904, making them the most modern ships in any fleet. But all the *Borodino* class suffered from instability because of a high center of gravity, which was exacerbated by overloading. They were inherently in danger of capsizing. Some naval architects have regarded them, in retrospect, as being among the worst battleships ever built. At Tsushima, already listing from their wounds, the two ships could return only feeble fire. At 1830 hours, a Japanese salvo hit *Alexander III*, which veered sharply to port and then abruptly keeled over. Half an hour later, she sank; from her 825-man crew there was not one survivor.

As twilight set in and light began to fail, Togo decided to call off the gunnery duel and pursue Rozhestvensky's crippled fleet with his plentiful torpedo boats. They circled like prairie wolves bringing down a great buffalo, scurrying about at high speed and with all the precision of months of hard training for this one engagement. If one can compare anything so barbaric and murderous as the destruction of a great fleet to a musical performance, it was as if Togo were conducting a superbly rehearsed symphony, and all its components were coming together in a thunderously magnificent last movement. Four boats from his Eleventh Torpedo Boat Division closed in on the blazing and abandoned *Suvorov*, where the heroic midshipman von Kursell kept up his solitary fire from the one surviving small-caliber gun, attempting to deter the lupine torpedo boats. At about 1920, Rozhestvensky's brave flagship, hit by numerous torpedoes, finally capsized "in a thick cloud of yellow smoke." In all this mayhem, what deck officers like Semenov and Pakenham could not witness were the horrors down below on the stricken ships, as stokers and engineers were scalded to death by superheated steam from burst boilers, or—if they escaped that fate—trapped below in rising waters. From the gallant *Suvorov* there were only twenty survivors. Kursell was not one of them.*

Borodino too was burning fiercely, but its crew was managing to drive off the torpedo boats with her 75 mm and 47 mm quick-firing guns. At 1928 hours, however, as Togo signaled to his battleships to cease fire with their main armaments, one of his commanders, Captain Matsumoto Kazu of the British-built *Fuji* (one of the older Japanese battleships, commissioned in 1897), decided to let loose a last salvo that had already been loaded into his twelve-inch guns. That lucky parting shot struck *Borodino* just below a six-inch gun, detonating the ready ammunition. Fire spread rapidly to adjacent magazines, which exploded with monstrous force, blowing out the hull. According to Pakenham aboard the *Asahi*, it produced "the sensation of the day . . . an immense column of smoke, ruddied on its underside by the glare of the explosion and from the fire abaft, spurted to the height of her funnel tops." There was left only "a dense cloud that brooded over the place she had occupied." Hidden beneath the cloud, *Borodino* swiftly capsized, the third modern Russian battleship to do so that day. There was only one survivor: Seaman First Class Semyon Yushin, rescued from the sea twelve hours later.

In one short afternoon the Second Pacific Squadron had lost four

***Suvorov* had had a short existence, serving for less than nine months within the Imperial Russian Navy before being sunk at Tsushima. Completed only at the beginning of September 1904, she had had virtually no proper sea trials. As a result, her crew was newly assigned and largely inexperienced.

of the world's most up-to-date and powerful battleships. Rozhestvensky's achievement, his epic, record-breaking voyage, bringing a fleet of fifty ships, often with semimutinous crews, halfway around the world to fight a major battle, had ended tragically in one of history's worst naval disasters. As a fighting unit it now no longer existed.

But what of its leader, the unfortunate Zinovy Rozhestvensky? Before being transferred to the destroyer *Buiny*, he was last seen aboard his flagship "sprawled in the disabled six-inch gun turret, his head, wrapped round and round with a blood-stained towel, nodding on his chest." In the tumultuous seas it was a remarkable feat that such a badly wounded man, his body limp and heavy, could be transferred safely to a tossing small craft, one already overladen with two hundred survivors from the sunken *Oslyabya*. With a splinter of his skull embedded in his brain, the tough old sea dog cheated death, drifting in and out of a coma, not ceasing to give orders in his moments of consciousness. He instructed that, with his flag of command, he be transferred to the destroyer, *Bedovy*, one of the few surviving ships with sufficient coal to reach Vladivostok, whither he now ordered it to head. By burning every inch of her wooden fittings, *Bedovy* could make the remarkable speed of twenty-two knots. Nevertheless, on the morning of May 28, she was overtaken by Japanese destroyers and brought to a standstill.

In a scene more in keeping with Trafalgar, a Japanese officer, Lieutenant Ayiba, leaped aboard *Bedovy*, unsheathing his sword and declaring, "I am now in command of the ship." Semenov, the only member of the staff aboard the destroyer who could speak Japanese, explained that the Russian commander in chief was below, but in no condition to be disturbed. Ayiba apparently could not comprehend the extent of his capture: the fleet commander aboard so insignificant a craft! Nevertheless, efficiently and with notable dignity, *Bedovy*, bearing the stricken Rozhestvensky, was towed into the Japanese port of Sasebo on the morning of May 30.

Meanwhile, as Togo resupplied and reorganized his battle fleet that night, the action turned to one of flight, pursuit, and humiliating roundup of the isolated, disconnected Russian ships that individually were still trying to reach Vladivostok. Safety lay just three hundred miles away. The Japanese cruisers, destroyers, and torpedo boats took over as the carnage continued. First to fall prey was the second-line battleship *Navarin*.* Although she was third from last in the Russian line of battle on May 27, little is known of *Navarin*'s actions, as there would be few survivors. She was hit late in the day by large-caliber shells that caused serious flooding and forced the ship to stop for repairs.

Rashly, *Navarin* switched on her searchlights to illuminate the attackers savaging her, but this served only to make her a better target for the torpedo boats. They may have achieved one or two torpedo hits in the dark. *Navarin* managed to get under way again and damaged one torpedo boat badly enough that she sank later that night. Around 0200 on May 28, the battleship was attacked again by Togo's Fourth Destroyer Division, which had laid six strings of floating mines ahead of her. Two of these mines struck *Navarin*, which swiftly turned turtle and sank. Three survivors were found alive sixteen hours later. The rest of her crew of 674 officers and seamen went down with their ship.

Next to succumb was *Sissoy Veliky*, a single-class ship planned to be the first of sixteen oceangoing battleships. Sandwiched between the doomed *Navarin* and *Oslyabya*, she had survived the daytime artillery duel with Admiral Togo's forces, though badly hit. The crew managed to extinguish the resulting fires and prevent her magazines from exploding, but could not contain the flooding of the ship. During the night the Japanese destroyers scored a torpedo hit, blowing off her rudder. Unable to maintain way, her engines were shut down, and

*Launched on October 24, 1891, the sixty-fourth anniversary of the Battle of Navarino in the War of Greek Independence, but completed only in 1896.

she was forced to surrender to a Japanese armed merchantman. The abandoned ship sank at 10:05 a.m. on May 28. Fifty members of the crew perished.

Meanwhile, unbeknown to the semiconscious Rozhestvensky, just over the horizon his de facto successor in the command, the despised Admiral Nebogatov, was surrendering his even more despised Third Pacific Squadron, together with four more battleships. These included the fourth of the five *Borodino* class, *Oryol.** During the battle of May 27, the Japanese fleet had concentrated its efforts against the Second Pacific Squadron, and so Nebogatov's ships had escaped the fate of Rozhestvensky's battleships that first day. With Rozhestvensky seriously wounded and most of the Second Pacific Squadron's warships sunk or disabled, command automatically devolved to Nebogatov. With daybreak on the twenty-eighth, his squadron was still plodding toward Vladivostok and safety, its speed sorely limited by boiler breakdowns and battle damage. He was commanding from *Nikolai I*, which, though hit five times the previous day, was otherwise virtually intact. With him was left only *Oryol*, two old coastal defense battleships, *Senyavin* and *Apraxin*, the lightly armored cruiser *Izumrud*, and a miscellany of lesser ships. Another elderly coastal defense battleship, *Admiral Ushakov*, had been left behind in the night but not discovered by Togo's torpedo boat wolf packs. Acting independently, three cruisers, *Oleg*, *Aurora*,† and *Zhemchug*, had abandoned any attempt to reach Vladivostok and were instead heading south for the Philippines and interment.

Suddenly, by 0930 hours, Nebogatov realized that he was sur-

*Of this class, only *Slava* now remained. Completed in June 1905, she was too late to accompany Rozhestvensky to the Pacific.

†*Aurora*, which had been a source of the Dogger Bank trouble in the North Sea, would survive through to the First World War, when she allegedly fired the first shot in the October 1917 Bolshevik Revolution in Saint Petersburg. She still sits as a prize exhibit in the Neva.

rounded by Togo's battle fleet. During the night *Oryol* had managed to repair one of her twelve-inch gun turrets. This was the only weapon available to Nebogatov that possessed the range to strike the Japanese. So Togo settled down to a long-range gunfight, to sink the Russian survivors without further risk to himself. It was obvious to Nebogatov that the Japanese intended to keep their distance and shell the Russians at leisure; and they had the whole day before them. Realizing that his ships were no match for Togo's, and that the Russian cruiser division under Admiral Oskar Enkvist would not arrive in time to prevent his destruction, Admiral Nebogatov decided to surrender at 1034 hours. He declared to his crews, in the knowledge that he might well be shot on returning home, "You are young, and it is you who will one day retrieve the honour and glory of the Russian Navy. The lives of the two thousand four hundred men in these ships are more important than mine."

Nebogatov then ran up the "XGE" flags: the international signal of surrender. For a while Togo would not accept the appeal to surrender, as it was so completely at odds with all the principles of Bushido, and he ruthlessly continued his bombardment. "It really was," he recorded, "the strangest occurrence, and we were astonished, and somewhat disappointed." Eventually he accepted the surrender, and four of Russia's surviving big-gun battleships sailed into Japanese ports.

The decision was not, however, accepted by all Nebogatov's officers, and Captain Vasily Fersen of the cruiser *Izumrud* disobeyed orders to escape through the Japanese lines. Unable quite to reach Vladivostok, its fuel running out, *Izumrud* was run aground just inside Siberian territory. The battleship *Admiral Ushakov*, having become lost during the night, was unaware of the orders to surrender and was caught by the Japanese the next morning, outgunned and outnumbered. She fought back gallantly, sustaining heavy casualties and being reduced to a blazing wreck, until her captain, V. N. Miklouho-Maklay, ordered her to be scuttled. Like so many of her proud companions, she capsized, taking Captain Maklay with her, though 328 of her crew were rescued.

Thus perished the last of the capital ships of the First, Second, and Third Russian Pacific Squadrons, and with it any hope of winning—or, rather, of not losing—the war. In this, history's most decisive of naval battles, Russia had lost all eight battleships that had set out under Rozhestvensky; twenty-one Russian vessels had been sunk, seven captured, and a further six disarmed or interned. Nearly five thousand Russian sailors had been killed, and almost nine thousand captured or interned, many of them wounded. And added to this were ships lost at Port Arthur or in the Battle of the Yellow Sea. In consequence, the tsar had lost not just all three squadrons of his Pacific Fleet; for all intents and purposes, he had in fact lost virtually his entire navy. It would never recover from Tsushima. From being the world's third-strongest fleet at the beginning of 1904, it now became a shaky sixth. And this great victory in the Straits of Tsushima had been achieved at the cost of just three Japanese torpedo boats, 117 dead, and 583 wounded.

Because of the bravery they had shown, the survivors of the *Admiral Ushakov* were treated by their captors with special consideration. When the stubby Nebogatov flew his white flag, he was said to have resembled a bent old bearded dwarf; the defeated crew lined up on deck in dirty and ragged uniforms, "like herds of tired grey sheep calmly awaiting their fate," in the words of one writer. Nebogatov too was received by the victors with samurai courtesy, and he and his officers were permitted to keep their swords. It was a tradition of Japanese chivalry, and humanity, that would die with Togo; certainly it would not be accorded to the defeated foe by the 1940s. The badly wounded Rozhestvensky, now a prisoner of war in a Japanese hospital, was treated with equal courtesy. Still lingering on the brink of consciousness, he remained unaware that Nebogatov had surrendered, and that his remaining ships were flying Japanese flags in the naval basin below. A Russian surgeon succeeded in removing the fragments of skull from his brain, and a few days later, as he regained his senses, the victor of Tsushima, Admiral Heihachiro Togo himself, came to visit him in the hospital.

Togo comforted Rozhestvensky with kindly words. He apologized for the somewhat spartan conditions in the hospital and "the absence of comforts due to such a distinguished patient." There was no need, he assured the defeated Russian, "to associate an honourable defeat with shame. . . . Defeat is a common fate of a soldier. There is nothing to be ashamed of in it. The only question is whether or not we have performed our duty." Togo praised the courage of Rozhestvensky's men: they had fought "most gallantly and I admire them all and you in particular. You performed your great task heroically until you were incapacitated. I pay you my highest respects."

Togo himself seems to have been taken aback by the completeness of his own victory, reflecting in his memoirs how his battleships had taken "years of labour to design and build, and yet they were only used for half an hour of decisive battle." Although that may represent some diminution of the facts, that half hour had certainly been *the* decisive half hour—in fact, one of the most decisive in all naval history.

Accompanying the Japanese commander in chief at this remarkable bedside visit to the vanquished Russian admiral was a young officer, a flag lieutenant called Isoroku Yamamoto. He had been wounded in the battle aboard the cruiser *Nisshin*, losing two fingers on his left hand, whether by enemy shellfire or by the breech explosion of one of the ship's own guns was never quite clear. Nor does any record remain of what Yamamoto's thoughts may have been that day; but, given the fame that was to descend on him little more than a generation later, they can well be imagined. Here he was alongside the fallen commander of the navy of the biggest nation on earth now at the feet of this small, gentle-mannered Japanese. How had it come about? A tiny island nation defeating such a monstrous great power, both on land

and at sea? Surely there was no limit to what such a nation manifestly blessed by the gods as Japan could go on to accomplish.

Rozhestvensky evidently responded with great dignity. In the only press interview he permitted, to a French reporter, he observed, "During the first half-hour our men fought well." But then they "suddenly became demoralized by the terrifying effects of the Japanese fire, and all was then over. . . . I was literally enveloped in flames."

Rozhestvensky made no attempt at self-justification, nor offered any excuses, declining to respond to a savage attack on him in the Russian press by one of his own former officers, Captain Nicolas Klado. On June 8 the tsar reassuringly telegraphed Rozhestvensky, "heartily" thanking him "and all the members of your squadron who have loyally fulfilled your duty in battle to Russia and myself." He wished the wounded admiral "a speedy recovery. May God console all of us." But this was not how Rozhestvensky, or Nebogatov, would be treated upon their repatriation home. While still a prisoner of war, Nebogatov was dishonorably discharged by the Russian Admiralty. On his return to Russia, he and seventy-seven of his subordinate officers were arrested and taken before a court-martial in December 1906.

On August 28 Rozhestvensky was deemed fit enough to leave the hospital and begin the long voyage home. On the first leg of the journey, en route to Vladivostok, the admiral had a foretaste of what might lie ahead for him at home when prisoners of war, being repatriated like him, rioted and broke into his cabin, demanding vodka and threatening "the man who spilt our blood." Back in Saint Petersburg, he was summoned before two courts of inquiry, where he took full blame for the defeat. He was honorably acquitted and retired on a generous pension. However, he insisted on appearing at subsequent hearings, first as a witness, then alongside the accused. Nebogatov and three ships' captains, including the commander of the destroyer *Bedovy* to which Rozhestvensky had transferred his flag, were sentenced to death by firing squad on December 25, 1906. The sentences, however, were subsequently commuted to ten years in prison by order of Tsar Nicholas. Nebogatov was released from the Peter and Paul Fortress in Saint Petersburg in May 1909 on the occasion of the tsar's birthday. All the others served long terms of imprisonment.

Rozhestvensky lived another four years, in quiet retirement. There was a bizarre alarum in July 1908, when the Admiralty announced details of a requiem service to be held for him. It was perhaps a final bar in the hymn to the abysmal incompetence of the navy of the tsar—the man who, with his councilors, really should have been made to stand before the courts. The news was published far and wide in the European press; like Mark Twain on a similar occasion ("The report of my death was an exaggeration"), Rozhestvensky took some pleasure in retorting, "You mustn't kill me off before my time." He lived peaceably for another six months, mercifully spared the horrors of 1914, of which Tsushima and the whole Russo-Japanese War had been but a prelude.

In the absence of radio, let alone television or the Internet, news of the full extent of the catastrophe at Tsushima was slow to filter through to Russian centers of population. On May 30, 1905, a large, expectant crowd assembled on the quay at Vladivostok. The local press had led them to believe a great victory had taken place; and rumors ran around relating the death of Admiral Togo and the destruction of his fleet. Then later there crept in three surviving Russian vessels, the destroyers *Bravy* and *Grozny*, each of 350 tons only, and *Almaz*, described as a yacht converted into a light cruiser. They were all that was left of Rozhestvensky's proud armada. With them their crews brought horrifying tales of what had really happened, of blazing and shattered capital ships turning turtle, of seas full of drowning Russian sailors, and of a gravely wounded commander in chief taken into captivity.

When the full scale of the disaster reached European Russia, the streets burst forth once more into revolt, but this time with more ominous preparation than Bloody Sunday the previous winter. On June 27, the famous mutiny of the battleship *Potemkin* took place at Odessa—nominally a rebellion over rotten meat, but one greatly inflamed by the news of Tsushima. Seven officers, including the captain, were killed; *Potemkin* bombarded Odessa, and then sought asylum in Romania. Before it was quelled, the mutiny had spread to other units in the Russian Black Sea Fleet, now the only viable naval unit left to the tsar. The standard of revolution was now truly raised, in what was to be a kind of dummy run for 1917, though the embers would, momentarily, appear to die down as Russia entered a brief new era of remarkable, perhaps inexplicable, prosperity on the very eve of the Great War.

CHAPTER 6

Peace

AFTER TSUSHIMA, AND coupled with Kuropatkin's defeat at Mukden, the time had clearly come to give peace a chance. The Russians had lost everything, but the Japanese were also almost at the end of their tether. Their armies in Manchuria were down to their last reserves, while Russian reinforcements still kept funneling in down the Trans-Siberian Railway. At sea, Togo had all but run out of ammunition as well as stocks of good Welsh coal. The bores of his big guns were worn out and in need of replacement, tasks that the Japanese arms industry could not fulfill. More dangerously, such had been the cost of fifteen months of all-out war that Japan's economy was tottering. Had it gone on much longer, the government would have had to give up and sue for peace on unfavorable terms. The country was simply exhausted, though none of this could be gleaned from the singularly uninformative and rigorously controlled Japanese press. In contrast, the immensity of the Russian landmass, and its vast population, could swallow its losses—at least superficially, and for a while. Even the meddlesome kaiser was fearful that matters were getting out of hand.

A peacemaker was needed, a kind of honest broker, like Bismarck. Into the breach stepped the substantial figure of Theodore Roosevelt, the twenty-sixth president of the United States, who, at forty-two years old when he assumed office, was the youngest-ever occupant of the White House. Roosevelt and his nation would not be thanked by the Japanese for their intervention.

Though a personality not prominently associated with works of peace on the international scene, "Teddy" Roosevelt—"Bull Moose," colonel of the Rough Riders who had captured Cuba's San Juan Hill during the Spanish-American War—threw himself into the process with his usual ebullience. Born a puny child, his early years dogged by chronic asthma, which forced him to sleep propped up in an armchair, "TR" had turned himself, by sheer willpower as much as anything else, into the supreme man of action. Naturalist, explorer, big-game hunter, rancher, amateur boxer (which had recently cost him the sight of one eye), author, and soldier, he was also a serious thinker. His book *The Naval War of 1812*, published when he was twenty-four, established his professional reputation as a serious historian; and, a voracious reader, he thought deeply about history. TR also wrote numerous other books on hunting, the outdoors, and current political issues, as well as frontier history. The inventor, and practitioner, of the expression "speak softly and carry a big stick,"* Roosevelt once bellicosely remarked: "I should welcome almost any war, for I think this country needs one." He was the force behind the completion of the Panama Canal, and (in 1907) sent America's Great White Fleet on a world tour to demonstrate American power. He believed strongly that the doctrine of America's Manifest Destiny should now also be extended widely into the Pacific.

TR was undoubtedly one of the most remarkable, if not one of the greatest, men ever to occupy the White House. Yet as an international negotiator he was an unknown quantity. Inside the elaborate French Second Empire–style buildings of the old State Department Building, with its black-and-white-tiled corridors (now the Eisenhower Ex-

*Roosevelt, then vice president, first used the phrase in a speech at the Minnesota State Fair on September 2, 1901, twelve days before the assassination of President William McKinley, which thrust him into the presidency. Roosevelt liked to describe his style of foreign policy as "the exercise of intelligent forethought and of decisive action sufficiently far in advance of any likely crisis."

ecutive Offices of the White House), he had a secretary of state, the quiet and gentle John Hay. But Hay was quietly dying and happy to hand over control to the president. TR greeted this opportunity with characteristic enthusiasm, uniquely taking sole charge of US foreign policy.

For ten months Roosevelt had little doubt that Japan would win the war against the bullying bear, an outcome he preferred. Following Togo's victory, somewhat undiplomatically he had congratulated his friend and fellow Harvardian Baron Kentaro Kaneko on May 31, 1905: "Every Japanese, but perhaps above all every Japanese naval man, must feel as if he was treading on air today."* Shortly after Roosevelt's inauguration in March 1905, Mukden had fallen, with terrible losses on both sides, and with Marshal Oyama urging on his government the need for an early peace. At the same time, Roosevelt began dropping subtle hints to the warring parties that the United States might be willing to offer "helpful offices" in the quest for an end to the war. But any formal request that it do so would have to come from them. He kept his hand closely concealed. Britain and all the European powers, Roosevelt felt, had their own special agendas that would be affected by the outcome of the war. Simultaneously there was a looming crisis between France and the kaiser over Morocco. TR then took off to hunt black bears in the remotest corner of the Colorado Rockies, leaving in charge his close and trusted associate William H. Taft, the portly secretary of war. "Camp Roosevelt" was cut off from the nearest telegraph office by thirty miles of snowdrifts and slush. There he awaited developments. Metternich could not have played it better.

*Kaneko, the staunchly pro-American minister of justice, had been sent as special envoy from Japan to enlist American help in bringing the war to a speedy conclusion, and immediately made contact with his fellow Harvard alumnus Teddy Roosevelt. As he sailed from Japan he wrote in a poem: "Bravely I head to my faraway former playground." (Kaneko would be one of the few senior statesmen in Japan to speak out strongly against war with the United States, as late as 1941; he died a few months after Pearl Harbor.)

Late on the evening of April 25, 1905, by which time Russia's Admiral Rozhestvensky was reaching the end of his remarkable eighteen-thousand-mile journey and was but a month away from the *Götterdämmerung* at Tsushima, Roosevelt received a coded message from Taft. Quoting a message from the Japanese foreign minister, Baron Jutaro Komura, to his ambassador in Washington, Kogoro Takahira, it contained a clear indication that Japan was now "anxious to effect peace through you." When he received it, the president had shot his third black bear (noting that it weighed about as much as the well-favored Taft; but did he, one wonders, also have thoughts of the despotic Russian bear?). Roosevelt was just recovering from a bad bout of recrudescent malaria. Snowed in at Camp Roosevelt, he spent his convalescence reading. For relaxation, he turned to Pierre de La Gorce's *Histoire du Second Empire* (all seven volumes, which he had brought along with him to the Rockies), his favorite hunt terrier, Skip, snuggling for warmth on his lap. Meanwhile the Russians were still sulking, the fierce tsaritsa Alexandra vetoing any weakness being shown by her husband. Crossly, TR wrote to Hay: "The Tsar is a preposterous little creature . . . unable to make war, and he is now unable to make peace." In the forthcoming collision at sea, Roosevelt predicted with clarity of vision that, though Rozhestvensky was clearly superior in numbers, "my own belief is that Japanese superiority in morale and training will more than offset this."

On May 6, weather at last enabled the president, his dogs, and his huntsmen to come down from the Rockies. When news of Tsushima came in, even he was stunned by the completeness of Togo's victory. He loathed the tsarist regime, but his sympathy for ordinary Russians and their culture made Russia much more congenial to him than Japan. What he could never understand about the Russian, however, "is the way he will lie when he knows perfectly well that you know he is lying"—surely a paradox that has perplexed negotiators dealing with Russia throughout the ages. The threat of revolution in Russia that followed Tsushima imparted to Roosevelt an extra sense of urgency. He pushed both his newly appointed ambassador to Saint Petersburg (and another old Harvard friend), George von Lengerke Meyer, and the Russian ambassador in Washington, Count de Cassini, to apply more pressure on the tsar, calling on Cassini to convey "his frank opinion that the war was 'absolutely hopeless for Russia.' "*

It was the eve of the tsaritsa's birthday, and therefore the Emperor of All the Russias was incommunicado. Nevertheless, Ambassador Meyer persisted and went out to Tsarskoe Selo, where he read to the tsar President Roosevelt's grim forebodings of disaster if peace could not speedily be concluded. Meyer could hear the cheerful sounds of Romanov children playing somewhere upstairs. Skillfully he had kept up Roosevelt's pretense that no prior initiative had come from Tokyo. Finally, and at last, Nicholas II acquiesced, with the proviso that it "be absolutely secret as to my decision, should Japan decline."

It was not until June 10, by which time both belligerents (Japan being the first by two days) had accepted Roosevelt's formal "invitation," that word was released to the world. The press was amazed, lauding Roosevelt's "great tact, great foresight, and finesse"; the London *Morning Post* declared that "as a diplomatist Mr. Roosevelt is now entitled to take high rank." Far from waiting to be garlanded with laurels, however, the president had once again disappeared into the boondocks. This time it was to a dacha on the Rapidan River in Virginia, once a scene of Civil War fighting. There he fried eggs and steak for breakfast for himself and his entourage, working off the effects by axing down some trees. For ten days he kept, uncharacteristically, absolutely dumb. The correspondent of *The Times* of London complained that even the State Department didn't know what was

*Marquis de Capuzzuchi di Bologna, Count de Cassini, was a Russian of Italian lineage born in Trieste, where his father was Russian consul. When the marquis went to Russia, he filled a number of diplomatic missions for the tsar. His grandson would be Oleg Cassini, the famous New York couturier.

going on. The British ambassador, Sir Mortimer Durand, was one of the few whom TR kept in the picture, perhaps surprisingly, considering that he rated Sir Mortimer's intelligence as being of "about eight guinea-pig power."* Britain nevertheless was one of the top players on the international scene. Durand faithfully passed on his information to the British foreign secretary, Lord Lansdowne, an astute statesman who had been instrumental in forging the Anglo-Japanese alliance of 1902, as well as the Entente Cordiale with France in 1904. The wheels of diplomacy started to turn, given extra impetus in Russia by the shocking news of the mutiny of the Black Sea battleship *Potemkin.*

Roosevelt's ailing secretary of state, John Hay, died on July 1, 1905, leaving TR entirely on his own in the tricky world of international diplomacy. Finally the lock tumblers began to fall into place, as both combatants responded to Roosevelt's cajoling and bullying. The president wanted the negotiations to take place in the United States, under his aegis. But Washington, DC, in summer, before the age of the air conditioner, was out of the question, so he chose the small New Hampshire town of Portsmouth.

To Portsmouth the Russians sent Sergey Witte, generally regarded as their ablest political leader and a staunch opponent of the war, who had been the tsar's minister of finance. It was a curious choice, in that Witte was detested by both the tsar and the tsaritsa, but it proved to be a brilliant one for Russia. Another Russian of German origin, he was considered by the imperial court to be vulgar, cynical, arrogant, boastful, and totally lacking in the modern-day quality of charisma. But on arrival in the United States he set out shamelessly to woo American public opinion, ostentatiously flattering Roosevelt.

Alongside the oppressive, self-confident bulk of Witte (he was al-

*TR found the ambassador a flabby walking companion, his inability to keep pace shinning up obstacles in Washington's Rock Creek Park a harbinger perhaps of Britain's declining power.

most six foot six, and wide in proportion—so wide in fact that he could not easily fit into a New Hampshire bathtub), the Japanese representative Baron Komura cut an insignificant figure. With his droopy black moustache, he somehow appears in photographs as a defeated figure, while Ambassador Takahira, top hat in hand, looks like a deferential tradesman or a family retainer who has just been given the sack. Neither exactly conveyed the image of delegates from a victorious nation. But there were also extenuating reasons for the gloomy faces of the Japanese delegates: Takahira had just learned that, chronologically, the Japanese were recorded as having sued for peace two days ahead of the defeated enemy. This would signify a serious loss of face to report back to the emperor.

Komura became increasingly glum as reports were arriving that the Russians were sending four fresh divisions to Manchuria; a winter campaign might have to be faced there. But, at home, all that the people could see, fed by a blinkered press, was the vision of Togo's Nelsonian victory at Tsushima. Where were its fruits?

The Russians agreed that Japan should regain embattled Port Arthur, but on lease only; they would recognize Japanese predominance in Korea, they would withdraw from Manchuria, and they would transfer to Japan the southern portions of the Manchurian feeder lines of the Trans-Siberian Railway. But they resolutely rejected Japanese demands for any financial indemnity ("not one kopeck," insisted the tsar), or for the surrender of the northernmost island in the Japanese archipelago, the bleak territory of Sakhalin, in close proximity to the Siberian coast. These were demands that Tokyo had regarded as paramount. At least twice it looked as if the negotiations were going to collapse, and Witte urged the tsar to break them off. By the weekend of August 25, rumors ran around Portsmouth that the Russians were checking out of their hotel. That same weekend, Roosevelt shocked his countrymen by taking off to dive to the floor of Long Island Sound in a risky newfangled submarine suitably named *Plunger*. It seemed symbolic of the state of affairs at Portsmouth, but TR refused to accept defeat. Twice he intervened with a personal appeal to the tsar. He confided to his son Kermit at the time, "The Japanese ask too much but the Russians are ten times worse."

At the eleventh hour, however, on Tuesday, August 29, the Portsmouth Conference ended. Witte placed a sheet of paper on the table with Russia's final offer. Japan could have half of Sakhalin, but no indemnity. Tokyo could perhaps have been offered better terms a week previously. Reluctantly but impassively, Komura accepted. The war was over, and a peace treaty would be formally signed on September 5.

The Japanese delegates went home in a somber and anxious mood. Witte was credited with having negotiated brilliantly, as Russia lost little in the final settlement. For his efforts, Witte was made a count.* Yet the de facto loss of the war would speed the beginning of the end of imperial Russia. Mutiny and revolt were breaking out everywhere, fed by anger and resentment among the repatriated troops at the excessive bottlenecks on the Trans-Siberian Railway, which were causing months, not just weeks, of delay on their way home. At one point, mutinous troops and strikers actually controlled the railway, until a trainload of regular troops was sent out to restore order with the usual arrests, hangings, and shootings. Historians have subsequently wondered whether, had Russia decided to fight on, and had it won a semblance of victory in Manchuria, revolution in 1917 may have been staved off. By August 1905, against a Japan whose ground forces (and finances) were on the ropes, the Russians had already assembled a new army in Manchuria exceeding a million men, and were in every way better prepared for action than they ever had been before.

If there was an overall winner, it was clearly President Roosevelt.

*Just before the outbreak of the First World War, Witte urged that Russia stay out of the conflict. His warnings went disregarded.

From all sides came admiration for his cool and levelheaded achievement in imposing peace on the two angry combatants. His friend the historian and journalist Henry Adams praised him as "the best herder of Emperors since Napoleon." Even the tsar and Kaiser Wilhelm II applauded and thanked him. The mikado was more enigmatically formal in his congratulations. The following year TR would, most justly, be awarded the Nobel Peace Prize, aged only forty-seven, and one of only three sitting presidents to be so honored; he promptly gave the award money away to a charity—which he could ill afford to do.* He could even permit himself to indulge in a modicum of self-congratulation: "It's a mighty good thing for Russia . . . and a mighty good thing for Japan . . . and a mighty good thing for *me* too!"

It would not, however, be enough to bring him back to the White House in 1912, at a time when the Western world could well have benefited from his peacemaking abilities.

TR seems to have sensed the dangers ahead. As the Portsmouth Conference had progressed, so his enthusiasm for Japan had waned. To his good friend Cecil Spring Rice, the British diplomatist whom he had wanted to see as ambassador in Washington in place of the mutton-headed Sir Mortimer Durand (an appointment he attained in due course), he wrote an extraordinarily perceptive and prescient letter:

> I am not sure that the Japanese people draw any distinctions between the Russians and other foreigners, including ourselves. I have no doubt that they include all white men as being people who, as a whole, they dislike, and whose past arrogance they resent; and doubtless they believe their own yellow civilization to be better. . . . No one can foretell [Japan's] future attitude.

*In 2001, Roosevelt was posthumously awarded the Medal of Honor. He was the first and, to date, the only president of the United States to be awarded America's highest military honor, and the only person in history to receive both his nation's highest honor for military valor and the world's foremost prize for peace.

> We must, therefore, play our hand alone. . . . Germany and France for their own reasons are anxious to propitiate Russia, and of course care nothing whatever for our interests. England is inclined to be friendly to us and is inclined to support Japan against Russia, but she is pretty flabby and I am afraid to trust either the far sightedness or the tenacity of purpose of her statesmen; or indeed of her people.

Eight years earlier, addressing the US Navy War College, Roosevelt had posed a hypothetical question. "Japan makes demands on Hawaiian Islands. This country intervenes. What force will be necessary to uphold the intervention and how should it be employed?" He continued:

> In a dozen years the English, Americans and Germans, who now dread one another as rivals in the trade of the Pacific, will have each to dread the Japanese more than they do any other nation. . . . I believe that Japan will take its place as a great civilized power of a formidable type, and with motives and ways of thought which are not quite those of the powers of our own race. My own policy is perfectly simple, though I have not the slightest idea whether I can get my own country to follow it. I wish to see the United States treat the Japanese in a spirit of all possible courtesy and generosity and justice. . . . If we show that we regard the Japanese as an inferior and alien race, and try to treat them as we have treated the Chinese; and if at the same time we fail to keep our navy at the highest point of efficiency and size—then we shall invite disaster.

As one historic consequence of Portsmouth, and of TR's intervention, henceforth the United States, having previously been no more than a bit actor in the Pacific area, now became a major player. Critics would see it as the "New American Imperialism."

Only the victors, the Japanese delegates Komura and Takahira, left Portsmouth unhappy and fearful. Their fears were swiftly realized. At home there was an angry and unprecedented reaction to the terms of the treaty. Once again, as ten years previously, many Japanese felt that they had been deprived of their rightful spoils, and by connivance among the Western powers. Such was the rigidity of Japanese censorship that, in February 1904, the Japanese had not even been informed of the outbreak of war until the evening of the following day.

Consequently, coming in on international wires, the news of the terms of the Portsmouth Treaty provoked a correspondingly violent reaction—in temperatures of 96°F (35°C). There was a mass antipeace rally in Hibiya Park in Tokyo, and after some fiery speeches, participants marched toward the Imperial Palace. Swords were drawn and the police overwhelmed. By the time order was restored the following night, seventeen people had died. Dozens of police stations had been burned down, together with many churches, newspaper offices, and even trams. Over four hundred policemen and forty firemen and soldiers had been injured before martial law, and heavy rain, brought an end to the riots that were unprecedented in orderly Japan. The riots also led, a short while later, to the collapse of Taro Katsura's government, on January 7, 1906. A new milestone had been established in Japan's relations with the United States, and with the Western world as a whole. Nevertheless, two years later—through the door opened at Portsmouth—Japan annexed Korea, the dream of expansionist Hideyoshi all those centuries earlier.

Following the Portsmouth Treaty, there was angry grumbling in particular aboard the ships of the Imperial Japanese Navy at what was construed to be a diplomatic surrender. The shock of the riots was compounded a few days later by a tragic disaster that overtook Togo's valiant flagship *Mikasa*, three and a half months after the guns had fallen silent at Tsushima. Anchored safely in Sasebo, she was blasted by an immense explosion that erupted in the after magazine. She sank in shallow water, together with 339 of her crew, nearly three times the fleet's total loss at Tsushima.* The exact cause of the explosion remains a mystery. In some respects it resembled that of the USS *Maine* in February 1898, which provided a catalyst for the Spanish-American War. There were immediate suspicions that the explosion on the *Mikasa* was an act of sabotage in protest against the Portsmouth terms. A much more likely culprit, however, is either the crew's carelessness or the spontaneous combustion of decomposing nitrocellulose propellant.

Admiral Togo narrowly missed going down with his ship, having left the *Mikasa* only a few hours previously. Responding in kind to the chivalrousness with which he had been treated when lying wounded in a hospital in Japan, Rozhestvensky sent Togo a telegram of condolence. For Japan it was a blessing that the disaster had not happened during the war; without Togo, the consequence for Japanese morale would have been at least as disastrous as the loss of Vice Admiral Stepan Makarov had been for the Russians. From the hour of Tsushima, Togo had become the subject of hero worship, amounting almost to semideification in Japan, where he would be lauded as "the Nelson of the East"; for all his innate modesty, it was a title that delighted him. Honors were showered upon him across the globe, including the Order of Merit awarded (in 1906) by King Edward VII. In Japan he collected such flowery decorations as the Collar of the Supreme Order of the Chrysanthemum, an honor held only by Emperor Hirohito and another royal prince, and the Order of the Golden Kite (First Class). Sparked by Japanese victories, a nationalist-minded poet in Calcutta, known as "the Kipling of Bengal," composed these lines:

---

*Raised from the seabed the following year, *Mikasa* became a hallowed shrine, surviving the Second World War and now set in a magnificent park adjacent to a statue of Togo.

Japan! Thy magnanimity like wild fire spread,
The proudest European Powers thee now dread,
Thou amazed all, all nations, thee the world adores,
Magnanimous Japan! Thy praise like a torrent pours!

When he stepped down from command, very quietly as was his wont, Togo gave a pointed farewell address, larded with parallels to Nelson and Trafalgar. He ended with an admonitory glance to the future: "The ancients well said: 'Tighten your helmet strings in the hour of victory.' "

The hero-admiral lived on into the 1930s, taking no part in politics except to register his objection to the discriminatory London Naval Treaty. Nevertheless, the myth of Japanese invincibility, which had grown up around him, would lie at the heart of the spiraling new militarism. On his death in 1934 at the age of eighty-six, Togo was accorded a state funeral with all honors. A seven-stanza poem of adoration on his death extolled him thus:

You are the treasure of Japan
The glory of the world
The embodiment of *bushido*
And the flower of the Far East.

Very different was the fate of Marshal Nogi, Togo's fellow victor in the ground conquest of Port Arthur and, later, Manchuria. He had been deeply shocked by the losses that the troops under him had suffered in the prolonged assault on Port Arthur: some fifty-six thousand dead, including his two sons. These deaths weighed him down for the rest of his life. At the war's end he expressed the depth of his feeling in a poem:

As a leader of the Imperial Army, I took a million soldiers
hostage.

> The battle resulted in a mountain of dead bodies.
> I am so ashamed of facing their old fathers.
> A song of triumph? But how many men can return home?

After the emperor had denied him permission to commit seppuku in atonement for his shame, Nogi spent the remainder of his life, and most of his personal fortune, on hospitals for wounded soldiers and on memorial monuments erected around the country to commemorate those killed during the Russo-Japanese War. On the day of Emperor Meiji's death in September 1912, Nogi committed ritual suicide. A moment before, his wife pierced her throat with a short sword, the traditional way of suicide for a woman of high birth. As a reminder of the old Japanese samurai tradition, Nogi's seppuku immediately created a sensation, and it had a profound impact on contemporary writers. For the wider Japanese public, Nogi became a symbol of loyalty and sacrifice, and he was sanctified as a "war god," a *kami*—a mystical hero of a different order from Togo, but a mystical hero nevertheless.

In Japan, Togo's acceptance that he was "the reincarnation of Horatio Nelson" was to become the root of fatal self-delusion in Japan as nationalism and militarism took hold in the 1920s and 1930s. A kind of mystical messianism regarded the low-loss victory of Tsushima 1905 as glorifying the "new" Japan, where the national spirit, coupled with an illusion of invincibility at sea, encouraged leaders of both the navy and army to contemplate an expansion to the Asian mainland and the Pacific islands. Was there any foe divine Japan could not vanquish? It was to become a suicidally dangerous mythology, as lethal to Japan as Nogi's seppuku had been: a fateful display of hubris.

PART TWO

Nomonhan, 1939

CHAPTER 7

Japan Goes Sour

AFTER THE LABORIOUSLY concluded Treaty of Portsmouth, imperial Japan launched itself into the new century laden with battle honors, on land as at sea, but also with a bitter sense of deprivation, of denial of its proper place in the sun, for which it had fought so hard. It had few real friends, and it was brimming over with a hubris that would lead it to a fatal conclusion at Hiroshima four decades later. The hara-kiris of Marshal Nogi and his wife contained within them a dark omen. Already by the end of 1905 the world had become a much more dangerous place than it had been at the beginning of the previous year.

The peace that President Teddy Roosevelt had so triumphantly concluded at Portsmouth, New Hampshire, in 1905 had provoked bloody riots in Japan, where crowds chanted "Don't stop the war," revealing a bellicosity that might have shocked, or at least seemed alien to, Western observers. With TR, the United States, revealing new interests in the Pacific, had now come forward as a potential fresh challenger to imperial Japan. The way ahead was marked by a series of keynote treaties and mutual "understandings" over the next few years. In 1907, the so-called Gentlemen's Agreement was negotiated. The background was anti-Japanese discrimination in California, where a school board had ruled that children of Japanese descent would be sent to segregated schools. Through Roosevelt's interceding, Japan agreed to halt the flow of immigrant workers to the United States in exchange for the school board rescinding the order. Hence the immediate goal of the Gentlemen's Agreement was to reduce tensions between the two Pacific nations.

The Japanese government expressed, pointedly, its strong desire "to preserve the image of the Japanese people in the eyes of the world." But the agreement was never ratified by Congress. There followed a little-known but most significant treaty, the Root-Takahira Agreement, signed on November 30, 1908, between an outstanding US secretary of state, Elihu Root (whom TR had appointed following the death of John Hay), and Kogoro Takahira, the unhappy negotiator of the Portsmouth Treaty.*

This agreement called for both nations to respect each other's territorial rights in the Pacific, and the independence and territorial integrity of China. But although Root-Takahira discreetly marked an important milestone on the route of the United States to becoming a major player in East Asia, it did little to meaningfully curb Japanese expansion into China.

The greatest benefit Britain gained after Tsushima undoubtedly stemmed from the digested lessons of the battle. The notes so assiduously made by Captain Willie Pakenham from his deck chair aboard the battleship *Asahi* arrived swiftly on the desk of the redoubtable first sea lord, Admiral of the Fleet "Jacky" Fisher.

In a flash, Fisher appreciated that Tsushima rendered the capital ships of every fleet in the world obsolete. It was the monster twelve-inch (British-forged) guns that had sent the Russian fleet to the bottom of the Tsushima Straits in a matter of minutes. Henceforth, thought Fisher, the fate of nations would rest on the all-big-gun battleship. It was fortunate for Britain that he had come to this conclusion before the German admiral Alfred von Tirpitz and his boss Kaiser Wilhelm II. With exemplary speed, Fisher ordered the construction of the world's first truly modern battleship. HMS *Dreadnought* was laid

*Like TR, Root would be awarded the Nobel Peace Prize (1912).

down on October 2, 1905, only four months after Tsushima, and was commissioned on December 2, 1906. It was an all-time record for warship construction, and a triumph for Britain's shipbuilding industry. An entirely new design, *Dreadnought* would remain in essence the role model for all capital ships until their extinction in 1945.

Following Tsushima, tsarist Russia plummeted from being the third-greatest naval power in the world to a poor sixth place; from being a first-rate force, it had come to be designated as only third rate. By 1914 it had still not made up its stupendous losses of 1905, the nine battleships sunk during the war. The role of the Russian Black Sea Fleet in 1914–18 was to prove indeed a bad "black" joke throughout the war. It is a tragic story, one that would eventually result in the deaths of many hundreds of thousands of British, Anzac, Turkish, and Russian lives at Gallipoli in 1915. Was it really unavoidable?

Some historians do not regard alternative history, the what-ifs, as altogether respectable. But I feel there are occasions when it is right and proper for the historian to postulate how consequences might have been transformed had certain events taken a different course. For instance, what if, of those great Russian battleships of Rozhestvensky's, some of them the most modern vessels then afloat, even half of them had not been sunk at Tsushima? Might not even three or four of these powerful units, reinforcing the Russian Black Sea Fleet, have reversed the tide of events, persuading the shaky Turks not to enter the war on the German side? What an immense difference this could have made to the course of the war—those doughty fighters, the Turkish Army, not arrayed against the Allies! As it was, by 1914 the Russian navy had still not recovered from Rozhestvensky's cataclysmic defeat. And by 1918 Turkey would have lost its entire empire.

Again, consider the influence of those Russian reverses of 1904–5 upon the rulers of Germany. Ironically, though he had supported Russia, the defeated nation, Kaiser Wilhelm II had been one of the principal beneficiaries of the Russo-Japanese War. Without firing a shot, he had witnessed the destruction of the fleet of one of his most immediate potential enemies in Europe, and the humiliation of its army in defeat in far-off Manchuria, at the hands of those he had once denigrated as "little yellow monkeys." He was a vain and foolish man given to faulty speculation, but on the strength of such a lamentably poor showing, he must have asked if it would not be worth taking a risk in Europe, to launch a knockout blow against the French enemy in the west while the Russian behemoth was still floundering in its own incompetence and the mud of eastern Europe. How much better would Moltke's and Tirpitz's sturdy Teutons perform against the vaunted Russian steamroller, now shown to be so profoundly flawed?

Finally, suppose that the reverses in the Far East, the humiliating fall of Port Arthur, had not taken place. Bloody January, the Revolution of 1905, would most probably have been avoided in European Russia. And, without that dummy run, might not the world-shaking cataclysm of October 1917 also not have occurred? Supposing, despite the loss of its fleet and the devastating defeat of Kuropatkin's army at Mukden, Russia had—as some of its leaders recommended—still fought on, against a Japan then so close to exhaustion itself, as described earlier? Might not some last-minute national triumph of arms in distant Manchuria have muffled the drums of revolt in Saint Petersburg, forestalling the events of 1917? Or at least have postponed them until Germany was decisively defeated, and not just through a fragile armistice in the west? How might the course of history not have been affected by one, or all, of these what-ifs?

Of all the belligerent nations in the First World War, none came out of it better than Japan. In 1914, the deal with its long-standing ally, Britain, was that it was to be rewarded with all of the German possessions in the Pacific. Loyally it helped the Royal Navy clear German raiders from the area, and it sent an efficient flotilla of destroyers to protect Allied convoys in the Mediterranean. Both functions materially helped Britain achieve its number one role of defeating the German navy in the Atlantic and North Sea. At the war's end, Japan was indeed given Germany's island colonies in the Pacific—the Mariana, Caroline, and Marshall Islands—as well as its strategic treaty port on the coast of Tsingtao. (There the Japanese navy conducted, portentously, the world's first naval-launched air raids, from a seaplane carrier.) In 1915 Japan exploited its position to impose its far-reaching Twenty-One Demands on China; it was a warning of which way nationalist thoughts might turn in the postwar world. These demands were immediately opposed by an outraged United States. Britain, engaged in a life-and-death struggle in Flanders, could do little more than register a bleat of protest.

In 1918, following the Bolshevik revolution in Russia, Japan and the United States sent forces to Siberia to bolster the armies of the White (anticommunist) movement leader Admiral Alexander Kolchak against the Bolshevik Red Army. Landing at Vladivostok, the Imperial Japanese Army initially planned to send more than seventy thousand troops to occupy Siberia as far west as Lake Baikal. Once more, Japan met with strong opposition from the United States, and its troops were the last to be withdrawn, with considerable reluctance. There were dual motives in Tokyo for lingering: first, there were still vivid memories of the cost of achieving victory over Kuropatkin in Manchuria in 1905; second, there was now on this critical frontier of imperial Japan a new potential foe that, having just murdered the tsar and his whole family, was the declared enemy of all imperialism. It was a frontier that the Japanese army in particular would watch with extreme nervousness over the coming decades.

The Great War, however, had cost Japan, for all its strategically valuable acquisitions, just a couple of hundred dead—by far the smallest casualty list of any combatant, representing 0.2 percent of their forces deployed.* The wartime boom boosted Japan's shaky economy

*By comparison, Britain had suffered nearly one million dead, the United States 117,000, and Russia and Germany each somewhere in the neighborhood of two million.

enormously, transforming it from a debtor to a creditor nation. Exports quadrupled between 1913 and 1918. Large combines such as Mitsubishi rose to the top. The war and the acquisition of German territories on the coast of China would give Japan greatly advanced leverage there in years to come. Most significant, however, was its representative Saionji Kinmochi sitting alongside the Big Four (Lloyd George, Orlando, Wilson, Clemenceau) at the Paris Peace Conference at Versailles in 1919. Japan gained a permanent seat on the Council of the League of Nations, a place at the highest table. Despite its relatively small role in the First World War, it had emerged as a great power; nevertheless, the Western powers, its fellow victors, had rejected Japan's bid for a racial equality clause in the subsequent Treaty of Versailles. This would sting, as would continued racial prejudice in the United States against Japanese immigration to California. (Back in December 1910, Theodore Roosevelt had made the prevailing American standpoint abundantly clear in a letter to his former secretary of war, now president, William H. Taft: "Our vital interest is to keep the Japanese out of our country. . . . The vital interest of the Japanese, on the other hand, is in Manchuria and Korea . . . our interests in Manchuria are really unimportant, and not such that the American people would be content to run the slightest risk of collision about them."

Meanwhile, tiny as their populations were, those islands in Micronesia would swiftly feel the boot of Japanese occupation. They "afforded a living to Japanese shopkeepers, restaurateurs, and brothel proprietors," wrote one historian, but "none of this was as important as their potential strategic value." Above all they would provide vital pieces in the race to build the vast empire of 1942. The occupation of these minute specks of coral in the Pacific also gave the Japanese navy a rationale for vastly expanding its budget, to double that of the army's. Thus, as the "peace era" of the 1920s began, the navy would gain significant political influence over national affairs and would intensify its rivalry with the army, a development fraught with danger for the rest of the world. At the same time, the imagined threat of Soviet Russia in Manchuria (it was no more than imagined because in fact Lenin, and later Stalin, had their hands full coping with the implementation of the Soviet system at home) would give the army an excuse to thrust its needs constantly into the foreground.

The then–provisional president of China, Yuan Shikai, foretold events with clarity, saying, "Japan is going to take advantage of this war to gain control of China." As it proved, Japan would exploit the opportunity of the Great War to impose her Twenty-One Demands on China while the European powers were too busy to stop her. These demands included an insistence on Japan's railway and mining claims and access to harbors, bays, and islands along the Chinese coast. In Barbara Tuchman's phrase, the Japanese made "incursions into Chinese sovereignty and territory which were to twist the history of the twentieth century." Yuan Shikai, in a weak position internally as his power was challenged by other local warlords, and unsupported abroad by the distracted Allied powers, was unable to risk conflict with Japan. So he accepted appeasement, a tactic followed by his successors, and agreed to an abbreviated document, the Thirteen Demands. These formed the basis of a treaty that was signed by both parties on May 25, 1915. Thus abbreviated, the treaty gave Japan few advantages that it did not already possess. On the other hand, the United States reacted strongly to Japan's rejection of the Open Door Policy, which had promoted equal trading rights in China for all nations. Japan's closest ally at that time, Great Britain, also expressed concern over what was perceived as Japan's overbearing, bullying approach to diplomacy, and the British Foreign Office in particular was unhappy with Japanese attempts to establish what would effectively be a Japanese protectorate over all of China. With rare prescience, the Foreign Office noted in 1921 that "Japan stands little chance of survival unless she can obtain control over the resources of China."

In China, the overall political impact of Japan's actions created much public ill will toward Tokyo, resulting in a significant upsurge in nationalism. A large-scale boycott of Japanese goods would be just one consequence.

Japan's wartime boom dissipated swiftly, leading to rapid inflation. In August 1918, rice riots caused by this inflation erupted in towns and cities throughout Japan. These disturbances demonstrated just how fragile the Japanese economy remained in essence. There was an unexpected rebellion in Korea in 1919; it was crushed, and at least seven thousand protesters were massacred, many of them students. This was so shocking that even the Japanese government acknowledged that something had gone badly wrong.

Nevertheless, Japan experienced moments of exhilaration and liberalism in the early 1920s, as did all the warring countries jubilant in the new peace. Down Tokyo's Europeanized center, the Ginza, long-haired young men in bell-bottom trousers, colored shirts, and floppy ties were to be seen strolling "with young women in bobbed hairdos." There were the "Marx Boys," the "Mobos" ("modern boys"), and "Mogas" ("modern girls"), the latter two often found tripping down the Ginza, hand in hand. Postwar Tokyo was marked by the trappings of "a skittish, sometimes nihilistic hedonism that brings Weimar Berlin to mind," as the historian Ian Buruma has observed; "on the whole, then, it was a good time to be young." By stern, old-fashioned Meiji standards this mood was very daring; the elders tut-tutted, but it prevailed, at least for a while. To Japanese historians this was the era of *ero, guro, nansensu*—or "eroticism, grotesquerie and nonsense."

It was a time of lovers' suicide pacts, but it was also when the first women's university was established in Tokyo; there were moves toward universal suffrage and women's rights in general. A wealthy Anglophile who emulated upper-class Britain, Takaaki Kato, prime minister in the mid-1920s, who headed perhaps the most democratic regime in interwar Japan, managed to pass a law giving the vote to all men over twenty-five with a steady income. Yet only a week later, a conservative Peace Preservation Law made it illegal to participate in any form of organized opposition to the *kokutai*, or national polity.* In practice this meant that communists or radical socialists could face up to ten years in jail. So much for political freedom.

Forms of government in Japan in the 1920s and 1930s may seem strange, if not incomprehensible, to Western minds. At the top of the pyramid, of course, was the emperor, the mikado, who claimed divine descent. But was he a constitutional ruler, like the sovereign of Great Britain; or was he an autocrat unique in his own right? On the death of the great Meiji Emperor in 1912, followed by a wave of ritual suicides, had come the Emperor Taisho. Of a weak physique, he died fourteen years later, in 1926. However, it was during his reign, known as the Taisho Democracy, that political power moved modestly from the oligarchy of ageing statesmen, the genro, to the Diet and democratic parties. But this brief bout of liberalism faded on the death of Taisho. His successor, the twenty-five-year-old Hirohito, 124th in the direct lineage of mikados, was academic by inclination, interested in natural science and marine biology. To stiffen him up, he was given regular tutorial sessions by the venerable Admiral Togo, the victor of Tsushima, who remained an influential figure in navy politics until his death. Under the Japanese constitution, the emperor could act only on the advice of his ministers and the chiefs of staff of the navy and army. The ministers of the armed forces did not, however, have direct access to the emperor; the admirals and generals did. He was so hedged in by protocol that, rather like the Oracle at Delphi, it was often not clear what the emperor's actual wishes were. Dangerously, the army and navy commanders with their direct routes of access could progressively interpret the emperor's remarks as they found most convenient.

In common with Europe and America, Japan was hit by the world

---

**Kokutai*, a difficult term to translate, was a semimystical Japanese concept that had something in common with Rousseau's philosophy of the General Will.

depression as the 1920s progressed. For small farmers, life became grim to the point of starvation. The indigenous silk industry was savagely hit by tough US tariffs. The population had risen from forty-four million to fifty-six million by 1920. Land hunger was acute, and the road to emmigration to America was barred. In 1923, Tokyo and Yokohama were devastated by the Great Earthquake. Perhaps a hundred thousand died. In the hysteria that followed, Koreans, communists, and anarchists were blamed for the natural phenomenon and attacked by mobs—and even by the police; mass killings took place. Yet within four years the two cities had been rebuilt, not least through generous aid from the United States. Nevertheless, out of the ashes arose a new kind of nationalism and imperialism. These ashes were fanned by the results of the Washington Naval Conference of 1922, the first disarmament conference ever to be held, but one invested with all the trimmings of old-style Western imperialism. Its aim was manifestly to impose a check upon Japanese naval expansion, much to America's advantage. Out of it emerged the 5:5:3 equation—meaning, in simplest terms, that the United States and Britain would each be allowed to build five new capital ships, but Japan only three.

In the eyes of many Japanese, this was no fair way to treat a wartime ally. Meanwhile, the Anglo-Japanese alliance of 1902, which had proved so profitable to both sides and was due for renewal in 1922, was allowed to expire. Britain now cast in its lot with the other Anglo-Saxon superpower, the United States—a switch that Washington's strategists saw as a way of ending British influence over Japan. Perhaps history would see this as an unfortunate development, insofar as it left Japan feeling, once again, dangerously isolated and bruised. Henceforth Britain's role as chief exporter to Japan of heavy machinery and weaponry dwindled. General Giichi Tanaka, who, of staunch samurai stock, had served as an aide to Kodama in the 1905 war and would later become a nationalist prime minister, castigated the Washington conference as "an attempt to oppress the non-Anglo-Saxon races, especially the coloured races, by the two English-speaking countries, Britain and the United States."

In 1920, California, followed by fifteen other states, passed tough alien laws, and two years later the United States Supreme Court ruled that Japanese nationals were ineligible for US citizenship. Ambassador Masanao Hanihara warned the US Congress not to exclude the Japanese in the proposed Immigration Act of 1924 because there would be "grave consequences." Senator Henry Cabot Lodge called this a "veiled threat," and the bill passed easily, causing considerable offense in Japan.

All this marked the rise of serious anti-Anglo-Saxon sentiments in Japan, which coincided with a period of political turbulence. In November 1921, Prime Minister Hara was murdered by a young right-winger, starting what would become a pattern in Japanese political life. In December, an anarchist fired a shot at the future emperor Hirohito but missed and was swiftly executed.

The confused and turbulent 1920s moved into the dangerous 1930s, the period that Japanese historians call the "dark valley." In 1930, the London Naval Conference allowed Japan to have 69.75 percent of what Britain and America were allowed in tonnage on heavy cruisers and 100 percent of their allowance on submarines. Captain Isoroku Yamamoto, he who, as a young flag lieutenant, had accompanied Admiral Togo to the bedside of the defeated Russian admiral after Tsushima, and now a champion of the Japanese navy's budding aircraft-carrier arm, forcefully ensured that there would be no limitations on naval aircraft; Japan alone of the naval powers already had four experimental aircraft carriers. But after Prime Minister Hamaguchi had managed to limit the navy budget to 374 million yen, he was severely criticized by the nationalists, and in November 1930 he was shot and wounded near the same spot in the Tokyo railway station where Prime Minister Hara had been killed. Despite several major surgical operations, Hamaguchi died the following August. Sinister right-wing, nationalist secret societies, such as the Cherry Blossom

Society (Sakurakai), the One Evening Society (Issekikai, formed in 1929), and the Black Dragons, began to proliferate. Under the inspiration of a fanatic called Ikki Kita, regarded as the founder of Japanese fascism, a group of politically dissatisfied young army officers came together to form a secret society with the objective of launching a coup d'état at some suitable future date, in order to combine the power of the emperor with that of the armed forces.

The curious thing about these conspiratorial Japanese factions, though they could indeed be loosely termed fascist, is that, in contrast to European experience, they never grew to comprise any mass organization such as Hitler or Mussolini had at their disposal. "What is more," notes the historian W. G. Beasley, "their avowed function was not to seize power but to destroy it. Out of chaos, they believed, would come in some mysterious way a better Japan."

In all this, a central and vaguely sinister role was played by the "Shinto myth," defined as a mystical bond, unlike anything to be found elsewhere in the world. In May 1932, the seventy-seven-year-old prime minister, Japan's twenty-ninth, Tsuyoshi Inukai, was gunned down by one of these gangs, comprised of eleven navy fanatics, in his own home. The trial that followed led merely to light sentences, which weakened not only the rule of law but also, in the long term, the young Japanese democracy. Inukai's death effectively marked the end of civilian political control until after the Second World War. Of the fourteen prime ministers who governed Japan between 1932 and 1945, only four were civilians. As this violent decade of economic stress unraveled, Japanese politics would be seen to be run not by respectable political parties, but by cabals of courtiers, military caciques, and bureaucrats, whose decisions, Ian Buruma has written, "were often forced on them by fanatical subordinates."

But it was abroad, in the inhospitable territories of northern Korea and southern Manchuria, and in the leased areas around Port Arthur, won during the 1905 war and designated "Kwantung," where the real roots of power resided. Here sat Japan's Kwantung Army. Nominally, its original role was to guard the railway line and fixtures of the South Manchuria Railway, which terminated in Mukden (or Shenyang today), against local banditry.* It was, however, gradually to grow into the most powerful unit of the whole Imperial Japanese Army and become a law unto itself. Its command was the most prestigious in the officer corps. As it developed, the Kwantung Army became not unlike those self-contained and self-ruling legions of ancient Rome, deployed to hold the marches of empire in its farthest corners, keeping the barbarians at bay, but also capable of overthrowing the home rulers themselves. The harsh remoteness of the bleak Manchurian steppes had its influence on those posted there. It was an ideal breeding ground for the perversion of the old principles of samurai and Bushido, with which Japanese militarism of the 1930s now clad itself. The troops, mostly of peasant origin and led by officers of the poorer middle class, were subjected to brutal discipline. They in turn would achieve levels of power and authority beyond their reach in the homeland. Senior officers and NCOs had served under heroes like Kodama and Nogi in 1905. Accordingly, they imposed on their subordinates exaggerated notions of invincibility, while burnishing the legend of the old hero Admiral Togo.

The Kwantung Army was pushed in two directions: one was a fear of the Russians to the west, now resurgent under Soviet rule and eager to requite the humiliations of the past; the second was a terrible racial contempt for the Chinese to the south. It was a contempt, deeply ingrained, that far exceeded any perceived slights by Anglo-Saxons toward the Japanese. Indeed, it even exceeded that of Nazis for the Slavs, and within only a few years it would overflow in the shape of

*Once a pioneer frontier town, Mukden/Shenyang now boasts an urban population of nearly six million, of whom 91 percent are imported Han Chinese, far outnumbering the native Manchus (who once ruled imperial China). Shenyang is now by far the biggest industrial city in northern China.

monstrous genocidal brutality and cruelty. Invisible to the West, the Kwantung Army was itself divided into two powerful factions, calling themselves the Kodo-ha ("Imperial Way School") and the Tosei-ha ("Control School"). Both were equally expansionist and ultranationalist. The noisier of the two, containing many young officers who favored a regime change by coup d'état, was the Kodo-ha faction, which was obsessed with thoughts of war against the Soviet Union in Siberia. A first step in that direction would be the occupation of all Manchuria. In contrast, the Tosei-ha plotters thought it prudent to stay on friendly relations with the Russians, and to make China the main objective of expansion once a base in Manchuria had been secured. Thus, both factions of the Kwantung Army were committed to an early occupation of all Manchuria, and to a program of aggression on the Asian mainland.

In September 1931, Seishiro Itagaki, the Kwantung Army's chief of intelligence, and Colonel Kanji Ishiwara carried out a carefully plotted fake attack in Mukden. They blew up a railway line and blamed it on Chinese "terrorists." The tracks were in fact only slightly damaged; trains passed over them soon after the explosion. But the "Mukden Incident" was sufficient to provide a pretext for the well-prepared Kwantung Army to move in and occupy, within a matter of hours, all the key points across the large territory of Manchuria.

Tokyo was not entirely surprised—rumors of a coup had been rife for some time—and the minister of war dispatched a major general to Kwantung general headquarters with orders to forestall the coup. But, illustrative of how much support the activists in Manchuria had at home, the general concerned went in a leisurely fashion by train to Mukden, where he allowed himself to be regaled in a geisha house by one of the plotters. By the time his letter had been delivered, on September 19, the coup had been carried out. In violation of orders from Tokyo, the Kwantung Army's commander in chief, General Shigeru Honjo, ordered his forces to expedite operations all along the South

Manchuria Railway. His troops captured virtually every city along its 730-mile length in a matter of days. Thus the whole of the vast, rich province of Manchuria—covering more than the total land area of France and Germany combined—was occupied with lightning speed, and now became Japan's puppet state of Manchukuo. Industrialists and land-poor peasants alike poured in. Manchuria also offered a vital springboard for an assault on either China or Russian Siberia. Both were in the minds of the Kwantung plotters. With great foresight, the Kwantung Army replaced the Russian wide-gauge railway track with its own of a narrower gauge. This meant that, in the event of a broader conflict, and any Soviet invasion of Manchukuo, the Russians would be unable to supply their forces there from the Trans-Siberian Railway, while the Kwantung Army would have the reverse advantage.

Manchuria was the first, the precursor, of all the expansionist landgrabs of the future Axis powers. Given its immense size, its acquisition certainly promised the space and the basic resources that Japan claimed it needed. Could the voracious appetites of Japan's industries and the land hunger of a soaring population be satisfied only by conquest and expansion? (One cannot help comparing the sixty-five million of the 1930s, when Japan already had access to the overseas territories of Taiwan and Korea, with a prosperous modern Japan, shorn of all its possessions outside the home islands, yet with a population almost twice as large.) The coup in Manchuria shocked the well-meaning mikado, his weak eyes immersed in the study of marine mollusks and bent on keeping out of everyday politics; he had not been consulted, though the hotheads of Kwantung claimed to have acted in his name. The Kwantung Army had committed an act of gross insubordination, but—in the face of success—Tokyo proved powerless to counter events.

Outrage swept the outside world, particularly in the United States. Nobody, however, was in a position to do anything. The nominal landlords of Manchuria, Generalissimo Chiang Kai-shek's nationalist

Kuomintang, who had ruled in Peking since 1928, were, as usual, in disarray at home; only adjacent Soviet Russia could have moved, but Stalin was likewise far too preoccupied with his own problems. From Geneva, the League of Nations protested. Japan declared this to be an unacceptable rebuke, and flounced out of the League in 1933, complaining that it was being victimized by the rest of the world. Its delegate, Yosuke Matsuoka, made a strange speech in which he compared Japan to Jesus Christ. A month later Hitler would come to power in Germany.

So Japan's Kwantung Army got away with a massive territorial acquisition, plotted by a few army colonels. It was an encouragement to Japanese nationalism for the future, though it meant that Japan now was isolated as never before, a dangerous situation that the Meiji Restoration had sought so diligently to avoid. During its fourteen years of occupation under the Japanese army's control, Manchukuo was one of the most brutally run regions in the world, and the greatest suffering was meted out to its despised Chinese inhabitants.

On February 26, 1936, Tokyo endured its heaviest snowfall in thirty years. Early that morning, young officers of the Kodo-ha faction launched a coup with over a thousand men, attempting to take over the center of Tokyo. Old, respected public figures were murdered. The leaders, none above the rank of captain, claimed that they were merely performing their duty as loyal subjects of the emperor. This was *gekokujo*, an inversion of the normal order, the lowly ruling over their lords. But Hirohito did not see it that way. It was mutiny, and the mild-mannered mikado, laying aside his microscope, now stamped his foot, ordering the rebels to be crushed. Martial law was imposed. The rebels gave up, and thirteen of them were executed, including Ikki Kita. Temporarily the Kodo-ha faction lost its sway; but, paradoxically, the political power of the army was strengthened, as well as that of the Tosei-ha, the hotheads bursting to move on China. Two weeks later, Hitler marched into the Rhineland; and at the end of the year Japan signed a pact with Germany, thus aligning itself with Hitler against the Soviet Union. The paths of aggression were mapped out.

On July 7, 1937, in the far southwest of Manchukuo, menacingly close to Peking, a Japanese private, Shimura Kikujiro, relieved himself beside the Marco Polo Bridge, then strolled off into the so-called demilitarized zone. The soldier was not away for very long, but since he was presumed to be missing on roll call, his commander insisted on combing the area for him. Private Kikujiro was later found unharmed. Meanwhile, however, there had developed a full-scale "incident," which was about as trumped up as the Mukden Incident of 1931. There was an exchange of fire; then, from July 9 onward, Japanese and Chinese violations of the cease-fire began to increase, and a buildup of reinforcements on both sides escalated; four divisions of Chinese troops moved to the border, and three Japanese. In charge of the Japanese troops was Kanji Ishiwara, now a general of the Kwantung Army. In Tokyo, the government of Prime Minister Fumimaro Konoe made statements threatening to step up Japan's mobilization, an even more aggressive move than the Kwantung Army could have hoped for. Meanwhile, in command of the Chinese forces was the new nationalist leader Generalissimo Chiang Kai-shek. Although debilitated through having to fight a war on two fronts—his men were also locked in a struggle with the communist forces of Mao Tse-tung—and poorly armed, Chiang's forces resisted the Japanese incursions with greater vigor than had been expected. Efforts to defuse the escalating conflict failed, largely thanks to the hard-line militarists within the Japanese General Staff. After launching a bloody attack on the Japanese lines on July 27, Chiang's forces were defeated and forced to retreat. The Japanese then moved in to pacify areas surrounding Peking. Chiang's access to the city seemed threatened, and negotiations failed.

Then, on August 9, 1937, a Japanese naval officer was shot in Shanghai. The war proper now began, a war that Japan would never be able to bring to a conclusion, with consequences that would shortly expand into a world conflict. In December 1937, Japanese forces, their strength swiftly increased to 200,000 men, seized Peking, Shanghai, and Chiang's provisional capital, Nanking. There followed the Nanking Massacre (December 1937–January 1938), in which an estimated 250,000 to 300,000 Chinese—men, women, and children—were slaughtered in one of the most shocking atrocities ever recorded. There was no trace here of the old-style samurai chivalry that had been shown the defeated Russians in 1905. In its place was reflected the years of vicious indoctrination in which Japanese soldiers had been told that the Chinese were an inferior race, the Japanese divine. In the years succeeding the Rape of Nanking could be seen the hand of hubris: the victors of 1905, followed by Manchuria, followed by Nanking, persuading themselves that they could now with equanimity take on the Soviet colossus as well as China—and what else besides?

Meanwhile, what was the rest of the world thinking as Japanese militarism rampaged, first across Manchuria, then into northern China? When, in 1933, Yosuke Matsuoka, sleek and sinister, and all in black, had stalked out of the League of Nations, the delegates, representing almost every nation on earth, sat stunned. "We are not coming back," declared Matsuoka as he left the chamber.* But in response to the Japanese incursions the League did nothing, its members declaring that they had neither the power nor the will. Britain and France were still licking their wounds from 1914–18, as was even the United States, tardy and modest though its contribution to that conflict had been. Soviet Russia, the one country that might have acted, was par-

*Matsuoka was the minister of foreign affairs up to the early days of the Second World War, and he was a fervent supporter of an attack on Siberia. He later admitted, however, that "entering into the Tripartite Pact [with Germany and Italy] was the mistake of my life. . . . I still keenly feel it. Even my death won't take away this feeling." He died in obscurity in an Allied prison in 1946 while awaiting trial on war crimes charges.

alyzed—as it had been in 1931—by purges and confusion at home. That first act of aggression by Japan would of course present Hitler with a perfect blueprint for the *Anschluss* (annexation) with Austria in 1938, and then the uncontested annexation of the Czech Sudetenland. If Neville Chamberlain could write off the dismembering of Czechoslovakia as "a quarrel in a faraway country between people of whom we know nothing," the violation of distant Manchuria would surely be even less troubling to Britain or the United States.

As for Japan and the Japanese themselves, the British view of their former allies showed a mix of supercilious derision, little short of racial contempt, and only moderate alarm. The prewar American view was probably little different. In Britain in the 1930s Japan was known for its gimcrack, cheap exports such as lacquered paper parasols that would last an afternoon. Books of the era reveal a steady increase in derisive anti-Japanese sentiment in the United States—not to mention all the anti-Japanese movies of the 1930s. They (the "Japs") had such bad, astigmatic eyes, it was claimed, that they couldn't possibly fly sophisticated planes. Aboard the battle cruiser *Repulse*, shortly before she was sunk by Japanese dive bombers, senior officers were heard to observe that "those Japs are bloody fools . . . can't fly . . . they can't see at night, and they're not well trained." As late as 1939, one American wrote, "The Japanese as a race have defects of the tubes of the inner ear, just as they are generally myopic. This gives them a defective sense of balance . . . the one physical sense in which an aviator is not permitted to be deficient."

Hence, the belief that the Japanese would simply never dare provoke a war with the United States was widespread. When the first Zero fighter attacked, senior Allied officers could not believe that the Japanese could independently design such an aircraft. In the Philippines, nine hours after Pearl Harbor, General MacArthur insisted that the pilots could not have been Japanese but must have been white mercenaries. One distinguished *New York Times* journalist, Otto Tolischus, would proclaim that the "Japanese mind works in a more elemental way . . . as a woman's is supposed to do." Even an erudite curator from the Smithsonian Institution in Washington could inform President Franklin D. Roosevelt, encouragingly, that "Japanese skulls are some two thousand years less developed than ours."

This supercilious attitude of 1930s Anglo-Saxons reminds one all too vividly of the contempt in which the Russians held the Japanese back in 1904. Understandably, it stoked up more resentment and hatred in Japan. Smugly, the British, distracted by Hitler's rampages in Europe, allowed their forces to sit without undue concern behind the giant coastal guns of Singapore, which would be a match even for the super 18.1-inch guns of the battleship *Yamato* (alas, they were pointing the wrong way: the Japanese would come by land, from behind, not by sea). The United States, with its more immediate Pacific interests, had harbored a much more critical view of Japan for some time, but had avoided taking any action. Then, on December 12, 1937, Japanese planes strafed, bombed, and sank a US river gunboat, the humble flat-bottomed *Panay*, anchored on the Yangtse River. Three men were killed, and forty-three sailors and five civilians wounded. Two newsreel cameramen aboard were able to film the attack. The survivors were later taken aboard a British gunboat, HMS *Ladybird*, which had herself been fired on earlier the same day. Eventually the Japanese apologized and paid the US government a handsome indemnity of $2,214,007.36. They claimed that their pilots had been unable to see the (very clearly painted) markings on the ship. US Navy cryptographers, however, had intercepted radio traffic that showed that the attacking planes were acting under orders and that there had been no mistake. The following month, an American diplomat, John M. Allison, the consul at the embassy in Nanking, was struck in the face by a Japanese soldier. For a moment it looked as if America might be on the verge of war with Japan. However, deeply isolationist at the time, with a large majority of its people determined to avoid any foreign entanglements whatsoever, and its forces at a low point, the United States turned the other cheek. Nevertheless, the two incidents turned US public opinion strongly against the Japanese.

CHAPTER 8

Zhukov, Sorge, Tsuji

ALL THROUGH THE 1930s the bellicose Kwantung Army would lead Japan further and further toward adopting its own aggressive stance. In 1939, a small border incident in Mongolia might seem to have little bearing on the colossal onslaught that would engulf Stalin's Soviet Union just two years later. But military adventures seldom happen in a vacuum; one leads to another, or they are otherwise interconnected, whether by events or by personalities. The Battle of Nomonhan, or Khalkhin Gol, as the Russians and their Mongolian allies called it, though it was fought thousands of miles away from Moscow, came to be laden with significance for the defense of the city that, in the autumn and winter of 1941, would prove to be one of the most decisive engagements in the whole of the Second World War.

Nomonhan would throw up two figures who would play a leading role in Hitler's defeat before Moscow, his first on land and an early turning point in the whole war. One was Georgy Zhukov, the unknown junior general later recognized as the most successful Allied commander of the whole war. At Nomonhan he would first try out the double-envelopment strategy that would later destroy the German Sixth Army at Stalingrad in 1942. The second important figure was Dr. Richard Sorge, Russia's ace spy in Tokyo, whose intelligence reports would enable Stalin to take gigantic risks in the struggle for Moscow. An earth-shaking change of strategy would emerge from Japan's humiliation on the field of Nomonhan: it would give up its ambitions in Soviet Siberia and instead "go south" in quest of plunder and raw materials, to the oil-rich East Indies, the Philippines, Singapore—and Pearl Harbor.

On June 1, 1939, as war clouds were building up over Europe, the forty-three-year-old middle-ranking Russian general Georgy Zhukov arrived in Moscow (restored as the Soviet capital in 1918). Responding to an urgent summons to the Kremlin, he had set forth with considerable trepidation from his post in Minsk, where he was a mere deputy to the commander in chief of Byelorussia (Belarus). Was he going to be charged with being an enemy of the people and end up in the Lubyanka prison, followed by speedy execution with a bullet in the back of the head? Stalin's Great Purge, in which many of Zhukov's generation of Soviet generals had met a grisly and generally unwarranted death, was just completing its terrible run. The dead included three marshals of the Soviet Union, headed by Mikhail Tukhachevsky (reckoned to have been one of the most outstanding military brains of his time), thirteen of Russia's fifteen army commanders, more than half the corps, divisional and brigade commanders, the commanders in chief of the navy and air force, and many other leaders of both these services. The numbers ran into thousands, if not hundreds of thousands. Exactly why the purges took place, no one can be quite sure even today; was it simply the Georgian tyrant's fear of an army coup against his brutal and incompetent rule? Suffice it to say that, as late as 1939, on the brink of a general world war, no senior officer in Stalin's Russia could feel safe.

What could be held against Tukhachevsky was that he came from the impoverished nobility. In this respect, Zhukov was unassailable: his proletarian antecedents were impeccable. He was born into a poverty-stricken peasant family in Kaluga a hundred miles southwest of Moscow, and everything about him indicated his peasant stock. In his teens, he had been sent to Moscow as an apprentice to a cobbler. During the purges Zhukov had attempted to keep his nose clean and avoid all involvement. With the bull neck, barrel chest, large hands, and somewhat coarse features of a Russian muzhik, he would become renowned for his ruthlessness, his unhesitating expenditure of lives, and his merciless punishment of failure.

Nevertheless, the summons—without explanation—to report to the people's commissar for defence, Kliment Voroshilov, in Moscow without delay would have filled him with anxiety. An old party hack, later described by those who knew him as a "bag of shit," the Ukrainian Voroshilov would be responsible for repeated military disasters, yet he was regarded by Stalin as a safe pair of hands. Notably, he had been his hatchet man during the purges, personally signing at least 185 execution lists.

Zhukov's concern was heightened when he reached Voroshilov's antechamber and was told by the aide to go in. "I'll see to it that all essentials are packed for you for the long journey," the aide added ominously.

"What long journey?"

"Go and see the People's Commissar. He'll tell you all about it."

After a routine inquiry about his health, Voroshilov informed Zhukov that "Japanese troops have made a surprise attack and crossed into friendly Mongolia which the Soviet government is committed to defend . . ." Showing Zhukov a map of the invasion area, Voroshilov continued: "I think they've started a big military gamble. At any rate, it's only the beginning. . . . Could you fly there right away and if need be assume command of the troops?"

Zhukov assured him: "I am ready this minute."

"Good. The plane will be waiting at the Central Airfield at 1600 hours. . . . A small group of military experts will be flying with you. Goodbye now, and all the best."

Zhukov departed for his new appointment with a mixture of relief, evidently, that his "long journey" was not going to take him to the dungeons of the Lubyanka and enthusiasm for the task ahead, which, he wrote in his memoirs, he approached with "a feeling of joy." Twice in his lifetime Russia had been humiliated by the Japanese in its Far Eastern province: once in 1905, and again during the Russian Civil War of the 1920s, when Japan had occupied eastern Siberia and had left only with the greatest reluctance in 1925—the last of the intervention forces to leave. It was clearly time that these combative Asians were disciplined. Given the threat from Hitler in the west, Zhukov appreciated that a limited operation to shut down provocations from Japan's puppet state of Manchukuo and secure the Soviet back door in the Far East assumed a very high priority.

A limited operation it did indeed seem to be. Once again the Kwantung Army had slipped its leash. To a tidy, rational Western mind its motivation might, to this day, have seemed irrational, if not inexplicable. The raw material wealth of eastern Siberia was largely unexploited. There was oil in the northern part of Sakhalin island, which Japan coveted, but nothing compared to that which was to be found in the islands of the East Indies to the south. The hotter heads of Kwantung hoped that maybe one day Soviet Russia in the east, a deeply flawed power, could be rolled back as far as Lake Baikal. Otherwise there was a deeply ingrained fear that the boundless commitment that Japan had taken on in China was menaced as long as its flank with Russia (which was then arming Chiang's Kuomintang forces) was insecure. So Soviet power in the area had to be neutralized, or perhaps taught a brisk lesson: a preventive strike, on a modest scale?

As we shall see in the next chapter, the Japanese had mounted an offensive in the disputed zone east of the Khalkhin Gol River, on the eastern border of Mongolia, in late May. Today the attack still seems politically questionable and militarily inept, and not with hindsight alone. East of Lake Baikal in Siberia, after all, the Soviets were believed to have between twenty and thirty infantry divisions. The Japanese forces seemed quite inadequate to the task. Yet it would be hard to think of a frontier that was riper for an "incident." On both sides there were forces that hated, distrusted, and despised each other. Only a generation previously vicious wounds had been inflicted at Port Arthur and Mukden. Since its coup de main in 1931, Japan had acquired a new frontier more than eighteen hundred miles long between the new puppet state of Manchukuo and its neighbors, the Soviet Union and Stalin's first satellite, the Mongolian People's Republic. Running through barren, sparsely populated land, much of the frontier was poorly demarcated in that originally the boundary lines had been internal divisions within the Manchu empire. Since the Russian Civil War there had been hundreds of border incidents, less or greater; one of these had come about when the Amur River had shifted its course. From Russia's point of view, its tenuous lifeline between Vladivostok and the western part of the country was greatly at risk; at one point, the frontier of Manchukuo came within only two and a half miles of a key bridge on the Trans-Siberian Railway.

Yet most of the incidents were initiated from the Japanese side. There was a congeries of right-wing politicians in Tokyo linked to activists within the Imperial Japanese Army—notably the Kwantung Army, with a Major Masanobu Tsuji much to the fore—who believed that the empire had certain inalienable rights in the Siberian hinterland that lay behind Manchukuo. Their memories reached back to the glorious triumph of Mukden in 1905 (without considering that it had been a very close-run thing, its heavy casualty list making it a pyrrhic victory indeed). They recalled how more recently, during the anarchy of the Russian Civil War, some seventy thousand Japanese troops had been the last to quit Siberian territory, shooed out by the United States. There were those who thought that all Siberia from Lake Baikal to Vladivostok, with its great empty spaces and largely untapped mineral wealth, should come under Japanese dominion. To them it was the Soviet empire, so rickety in appearance, that was Japan's natural enemy in the 1930s, its Marxist ideology fundamentally hostile to imperial Japan. Fueling the smoldering tensions in Soviet-Japanese relations was Japan's signing the Anti-Comintern Pact with Nazi Germany in November 1936.

In the summer of 1938 the most serious of the border incidents occurred over some disputed heights at Changkufeng, on the Tumen River that bounded Manchukuo to the east. From the Soviet point of view, Changkufeng was perilously close to Vladivostok. The Japanese incursion seems to have been triggered by the defection of a Soviet general, Genrikh Lyushkov, who had commanded the NKVD forces, the dreaded secret police, in the Soviet Far East. He was said to have provided the Japanese with important intelligence on both the poor state of Soviet Far Eastern forces and the continuing purges of senior army officers. It was a major intelligence coup for Japan, yet from it their analysts allowed themselves to draw exaggerated conclusions about the damage the purges had done to the Soviet forces facing them. The Changkufeng Incident escalated as Moscow mobilized the Pacific Fleet. The Japanese forces suffered a humiliating local defeat and, after ten days of bloody fighting, were pushed off Soviet-claimed territory. A cease-fire was negotiated, but not before the Russians had suffered eight hundred casualties, while the Japanese had lost approximately twice that number. The Russian commander, General Vasily Blyukher, was blamed for the excessive Soviet casualties and was executed. Both sides withdrew, bloodied and with hackles raised.

Inferiority in manpower, in weaponry, and in generalship would all contribute toward the Kwantung Army's coming failure at Nomonhan. Yet perhaps as telling as any of these was one factor that worked, undetected and unseen, to defeat Japanese arms: poor intelligence. As Alvin D. Coox puts it in his impressive study, "No one expected the Russians to become heavily involved in the Nomonhan fighting. . . . It is undeniable that Japanese underestimation of the Red Army aggravated the shortage of hard facts."

Even if the Kwantung Army had been of the very first order, nothing could stand up against the one ace in the pack held by the Kremlin. Whereas most of the British and American successes in the domain of intelligence in the Second World War came from intercepting signals (sigint), the Soviets held the advantage in human intelligence (humint), in material supplied by operators such as the Cambridge spies Burgess, Maclean, and Philby. But in terms of sheer value, none could approach the status of Moscow's man in Tokyo, Dr. Richard Sorge.

Born of a German father and a Russian mother in 1895, Sorge was badly wounded on the western front in 1916 and invalided out of the German army, decorated with the Iron Cross. While convalescent, he described himself as "plunged into an intense confusion of the soul" and joined the German Communist Party. In the 1920s he served an apprenticeship in Moscow, where he was recruited into the GRU (military intelligence). He then moved to Shanghai, working as the editor of a German news service and for the respectable *Frankfurter Zeitung* newspaper. There he set up an intelligence ring, in which one of his agents was Max Clausen, also a German Communist. In those early days he seems to have done remarkably little to cover up his communist sympathies. Then, in 1933, the year Hitler came to power, his GRU controllers ordered him to Tokyo to set up an important spy ring.

The network he created included Max Clausen as radio operator, and various other Comintern agents, among them Clausen's wife, Anna, who served as a courier. Hotsumi Ozaki, another Sorge recruit, became a member of Prime Minister Fumimaro Konoe's inner circle and would acquire much valuable intelligence from Kunoe, which Sorge would promptly dispatch to Moscow.* The GRU appears to have attached utmost importance to the establishment of Sorge's network.

*When the Sorge ring was broken up in 1944, Ozaki achieved the rare distinction of being the only Japanese national to be executed for treason in the Second World War.

Meanwhile, Sorge achieved a close friendship with Major General Eugen Ott, a military attaché who in April 1938 became the German ambassador in Tokyo and who confided in him as a devoted Nazi (Sorge had conspicuously joined the Tokyo branch of the Nazi Party in 1934). Ott seems to have told Sorge everything he, or Moscow, wanted to know. The intelligence was promptly sent to Moscow on unbreakable one-time pads. As Sorge was to admit in his later confession, his principal assignment in the 1930s was "to observe most closely Japan's policy towards the Soviet Union following the Manchurian Incident, and at the same time to study very carefully the question of whether or not Japan was planning to attack the Soviet Union." Sorge remained faithful to his creed: "to free the world from capitalist imperialism and militarism, and . . . make wars impossible in the future. All difficulties, shortages and mistakes in the Soviet Union were only regrettable transitional phenomena."

Though he seems to have left numerous traces of his past communist life, to none of his German friends in Japan did Sorge "ever betray by the slightest hint, or slip of the tongue, that he had firsthand knowledge of the Soviet Union." He visited Manchuria, and by 1935 he and his fellow spy Yotoku Miyagi were able to prepare an elaborate diagrammatic chart showing the leadership, alliances, and enmities of the rival factions in the Japanese army. Thus he was uniquely qualified to interpret the insurrection that hit Tokyo in February 1936. As Miyagi correctly deduced, "it would be a turning point, leading to a drastic redirection of Japanese politics, with the army becoming the motive power in political life." That, as we have seen, is precisely what happened.

Sorge warned Moscow of the likelihood of Japan's adventure in China, and in June 1938 the German embassy made him privy to information acquired from the interrogation of the defecting Soviet general Genrikh Lyushkov. A hundred percent loyal to Stalin, Sorge was profoundly shocked by what Lyushkov had revealed of the debility of the Soviet armed forces in the aftermath of the Great Purge. What particularly shook him was the temptation that such material could spark in the mind of Hitler, as the Soviet Union's number one potential enemy. As he remarked later: "One consequence of Lyushkov's report was a danger of joint Japanese-German military action against the Soviet Union." It was, observed his biographers F. W. Deakin and G. R. Storry, "Sorge's main task to contribute towards the avoidance of such a disaster."

It is not clear whether in 1939 Sorge tipped off Moscow about the impending Japanese attack at Nomonhan; there appears to have been a temporary breach in communications, possibly because Sorge's GRU controller was himself caught up in the purges. What is known, however, is that Sorge was able to assure Moscow that the Nomonhan Incident would not be allowed, by either the Japanese government or the army, to escalate "into real war." He was recorded as expounding to Ott during the battle that "if the Japanese Army wanted to drive the Red Army from its present positions, then 400–500 tanks would be required; and this was beyond Japan's industrial capacity. Germany ought to study the whole Nomonhan Incident more deeply and should reject the old idea that the Red Army was incapable of putting up serious resistance."

History would shortly reveal that, disastrously, Hitler chose to give credence to Lyushkov's report. As it was, Sorge's report about the limitations of Japan's objectives and its military capacity would prove of immeasurable value to Stalin in the formulation of his tightrope strategy in the summer of 1939. It would also have prescribed to him the size of Zhukov's forces that would be required to smash the Kwantung Army deployed at Nomonhan.

In April 1939, eight months after Changkufeng, Major Masanobu Tsuji promulgated a fiery Order 1488 at the Kwantung Army's headquarters. In it were contained "principles" for dealing with any future frontier clashes. These instructed that "if the enemy crosses the frontiers . . . annihilate him without delay." To accomplish this, it would be permissible therefore "to enter Soviet territory, or to trap or lure Soviet troops into Manchukuoan territory. . . . In such an event, however, never abandon friendly dead or wounded. . . . In the event of an armed clash, fight until victory is won regardless of relative strengths or of the location of the boundaries."

If "the enemy" were to "violate the borders," the order continued, Japanese units "must challenge him courageously . . . without concerning themselves about the consequences, which will be the responsibility of higher headquarters." It all indicated that the Japanese High Command, swallowing its own agitprop that the Soviet Army in the Far East was but a paper tiger, had refused to learn anything from Changkufeng: for instance, that the Soviet Army in Siberia was tougher, was better equipped in tanks and (especially) artillery, and had been significantly reinforced. The Japanese intelligence staffs would revise none of their low assessments of Soviet military power in the Far East. "If we do not seize an opportunity such as this to hit the Russians and thus show the power of the Imperial army, then the Soviet-Manchukuo border will hereafter be dominated by Soviet forces, and that will only leave roots for trouble in the future."

Major Tsuji can be taken as an avatar of Japanese aggression in Kwantung and afterward. Like the villain of some Victorian melodrama, he would pop up everywhere in the Pacific war zone where there was trouble, spreading the doctrine of Japanese racial superiority and the supreme honor of suicidal resistance. Born at the beginning of the century, Tsuji was thus brought up in that generation of Japanese steeped in the legends of the heroism of the 1904–5 war, that pyrrhic victory on land, for whom the self-immolating Marshal Nogi was the ultimate hero.

Tsuji published various writings that were read by a vast public in Japan. They are hymns of self-praise, but even if only 25 percent are trustworthy, they still reveal a phenomenal man of war, utterly fearless (and wounded many times), with a transcending genius for military planning and an irresistible ability to put his ideas across to senior commanders.

At the Kwantung Army's general headquarters, Tsuji's slot was simply that of senior operations staff officer, but his powers ranged much further. He was present at the Rape of Nanking (over which his writings draw a discreet veil). And he seems to have almost singlehandedly prepared the brilliant operational plan by which the Japanese Twenty-Fifth Army would infiltrate on bicycles down through the jungles of Malaya, ultimately to capture Singapore from the rear, forcing a humiliating surrender upon a British army of 130,000 men in February 1942.

Nicknamed the Wolf, or the God of Operations, he was also responsible for training a force in the jungles of Taiwan and southern China, which he led from the vanguard. After Malaya, Tsuji appears in the Philippines; then he turns up in Guadalcanal during 1942–43, planning—and leading—two final attempts by the hard-pressed Japanese defenders to dislodge the US Marines. From there he went as special emissary to the emperor to urge evacuation.

After that, Tsuji reappeared in Burma in the latter stages of the war. In every theater where he set foot, charges of brutality and war crimes emerged: the massacre of Chinese civilians in Singapore, the executions of US prisoners of war during the Bataan Death March and of government officials in the Philippines, and numerous killings in China. None, however, was more shocking than that with which he was charged in Burma. As the war historian Max Hastings writes, he was reported there to have dined on the liver of a shot-down American pilot, damning as cowards his comrades who refused to share it with him: "The more we eat, the brighter will burn the fire of our hatred for the enemy."

With the ending of the war in August 1945, Tsuji's career continued to be almost equally unorthodox. Though placed high on the

Allied wanted list of Japanese war criminals, and regarded as the most notorious of these to evade trial after the war, he managed somehow to escape by going into hiding in Thailand. Unconfirmed accounts had it that he was helped through collaboration with the CIA, who were then working with the rump of Chiang Kai-shek's regime in southern China. Once a moratorium on the war trials had been pronounced, Tsuji returned to Japan, where his memoirs became best-sellers and brought him renown.* He was elected as a member of the Diet; then, in 1961, he set off on his travels again, this time to war-torn Laos. There he disappeared, presumably caught up and killed in the civil war. In his hometown of Kaga a memorial statue, dressed in a formal tailcoat, stands to honor his memory. Sometimes one may wonder what kind of people would want to commemorate a man like Masanobu Tsuji.

The basic military ethos of apostles of *gekokujo* like Tsuji, which would propel Japanese forces all the way through the Pacific War, was simply stated. It was a repetition of Georges Danton's famous war cry from the French revolutionary wars—*"de l'audace, encore de l'audace, toujours de l'audace"*—and, coupled with that, the notion of "no surrender." Of the heroes of 1905, if it was to Admiral Togo whom Tsuji and his apostles looked to for the touchstone of daring, it was even more to Marshal Nogi whom they turned for lessons of self-sacrifice. These lessons would be repeatedly evoked in the 1930s as representing the soul of Divine Japan, the Will of the Emperor. Delving back into samurai glories of past centuries, the *philosophes* of the military brotherhood controlling the Kwantung Army latched on to the principle that death in battle was the noblest of all achievements, whereas to surrender a post or be taken prisoner was the worst imaginable

---

*Notably among his memoirs is his blood-and-thunder account of the fall of Singapore, *Japan's Greatest Victory, Britain's Worst Defeat* (New York, 1997; trans. Margaret E. Lake).

disgrace. There wasn't even a word in current military lexicon for retreat, fallback, or withdrawal; instead, commanders would speak of "moving in a different direction." Unheard of was that maneuver popular in the British Army, "falling back to prepared positions," let alone, as experienced in the American forces' worst moments in Korea in 1950, the "bug-out," or disorderly flight.

From Tsuji at Kwantung Army headquarters would come regular enjoinders that, in the event of conflicting orders, the recipient was always to obey that which was most aggressive. Young officers were constantly inculcated with the notion that, in battle, moral and spiritual strength would win out over material factors, a tenet deeply rooted in the 1905 war. Here was a philosophy that would bring Japan some astonishing and cheap victories during the early, aggressive stages of the Pacific War but would ultimately lead it into utter ruin against the material weight of the most potent economy the world had ever seen. And, more immediately, this philosophy would lead to a similarly decisive defeat—though it was visible more locally—in the forthcoming Nomonhan Incident.

Ever since the Changkufeng Incident, Tsuji and the Kwantung hotheads had been spoiling for a fight, unhappy with the alleged timidity of the higher headquarters and displeased with the doves in their midst. The Kwantung Army as a whole was resentful of the vacillation and interference of the General Staff in Tokyo.

Meanwhile, unnoted in Hsinking, the capital of Manchukuo, a particularly bellicose speech was delivered by Marshal Voroshilov marking October Revolution Day in Red Square on November 7, 1938. Referring to the recent fighting at Changkufeng, he warned that if another such border violation occurred, the Red Army would not confine itself merely to repelling the incursion: "It is more convenient and easier for us to crush the enemy in his own territory." In a report to Moscow, General Grigory Shtern, commanding all Soviet forces in the Far East, warned that although Changkufeng was only an episode, "it was a real operation of modern war for which the Japanese troops had undergone special training and the Imperial Staff had carried out detailed reconnaissance over many months." Moscow was beginning to take these small-time operations in its distant Far East with extreme seriousness.

Thus was the scene set for the next "incident," at Nomonhan.

CHAPTER 9

The Incident

THE BATTLE OF Nomonhan took place diagonally at the opposite end of Manchukuo to Changkufeng. There the easternmost parts of the Mongolian People's Republic (Stalin's first satellite) and Inner Mongolia (nominally Chinese) together formed a salient, shaped rather like a pig's snout, biting a chunk out of western Manchukuo. It was a miserably harsh, semiarid area of mostly flat, sometimes gently undulating, sandy plains and grassland studded with occasional short scrub pines and low shrubs. The climate was extreme; in May the days could be hot, with freezing nights. In July and August, the daytime temperature regularly rose above 100°F (38°C), but the nights remained bitter. Mosquitoes and enormous horseflies swarmed in the summer, making netting imperative. There was little rainfall, but in the summer, especially in August, dense fogs often rose at dawn. It was hard to see what strategic value the area might have to either side. But to Tsuji and his fellow planners it composed a vital flank wherewith to pursue the stalemated war with China, and equally a flank on which the Russian bear, if untamed, could thwart Japanese intentions.

The frontier in question was actually one of the few stretches where there could be, or should have been, a clear demarcation: the meandering Halha (or, in Russian, Khalkhin Gol), a modest river starting in the Greater Hsingan mountain range in southern Manchuria (Manchukuo) and flowing to the northwest to empty into

Lake Buyr Nuur. At Nomonhan the river was a meager fifty yards wide and roughly six feet deep, flowing over a sandy riverbed.

This was the border claimed, reasonably enough, by the Japanese for their new puppet state of Manchukuo. The Russians, however, on behalf of their Mongol clients, had drawn a shorter, arbitrary line some ten miles farther east, through the small village of Nomonhan. This claim was based on the fact that, from time immemorial, and unchecked through the Qing dynasty, Mongolian nomads had habitually taken their shaggy little ponies across the Halha to graze on the marginally less wretched pasture there. Then, one day in the spring of 1939, the ponies came across accompanied by armed Mongol cavalry. Shots were fired by Manchukuo frontier guards. On May 4, the Japanese observed a further party of more than fifty Mongolian horsemen setting up positions close to Nomonhan. A firefight ensued that lasted for ten hours and gave rise to several casualties. The situation began to escalate rapidly, each side giving very different accounts of who was responsible.

Although twenty-seven out of Japan's thirty-seven divisions were committed to the war in China at the time, a whole division, the Twenty-Third, was now dispatched to the desolate front at Nomonhan, as a token of how seriously conflict with Russia was taken. The Twenty-Third was a new formation, a "triangular" division based on three regiments only. Its troops had come from training in Japan's subtropical Kyushu Island, and the division was said to be "near the end of the queue" for equipment. It was therefore assigned to what was originally believed to be a quiet backwater. The division was short on machine guns and particularly deficient in antitank weapons—but was enemy armor likely to appear on the plains of western Manchuria? The Twenty-Third was commanded by a rather gentlemanly commander of the old school, the fifty-two-year-old Lieutenant General Michitaro Komatsubara. Considered to be one of the Imperial Army's leading Russian experts, Komatsubara had served two tours as a military attaché in the Soviet Union. He was solidly built, wore eyeglasses, and sported a small mustache. Meticulous in his habits, he kept detailed diaries, wrote long letters, and composed poems; but he had had no combat experience. Though he had an air of gentleness about him, he was described as also exuding "a sense of gloom." In common with all in his command, Komatsubara was steeped in the gospel according to Masanobu Tsuji, then a senior operations staff officer at the Kwantung Army's general headquarters. But given that Nomonhan was such a remote and unimportant location, not one of the Kwantung Army's headquarters officers knew where it was; the Twenty-Third Division's staff had difficulty locating it themselves.

Komatsubara's Twenty-Third Division, entirely reliant on the Kwantung Army for information, says Alvin Coox, "knew nothing about enemy forces behind the river line." Later, staff officers at Hsinking admitted that the Japanese strength would have been insufficient to annihilate hostile forces in the Halha region, "since those forces proved to be far larger than had been anticipated at the time." They knew that the nearest Soviet railhead from which remote Nomonhan could be reinforced was at Borzya, some four hundred miles away, and was itself on a single-track spur of the Trans-Siberian Railway leading from Chita and Lake Baikal. In contrast, the Twenty-Third Division could be swiftly supplied or reinforced over Manchukuo's well-developed network of railways, the railhead at Handagai just fifty miles distant. Little was known about the Soviets' strength in hand, or their potential. But did that matter? Were they not the same torpid Russian enemy, a backward race that Nogi and the parent generation to Tsuji and Komatsubara had drubbed so decisively just three decades earlier, and were now cowed by a generation of communist muddle, brutality, and mindless purges?

So, sometime in mid-May 1939, a troop of forty Mongolian cavalry crossed the Halha River and established a camp on a dusty knoll at Nomonhan—thus well inside what the Japanese regarded as being their territory. A Manchukuoan cavalry unit was dispatched to see them off. There was an exchange of fire, resulting in casualties. Each side reported the incident to its senior formation. The battle, or war, of Nomonhan, had begun. General Komatsubara, mindful of Tsuji's Order 1488, reacted strongly, sending forward a whole battalion of his reconnaissance regiment under Lieutenant Colonel Yaozo Azuma, some 220 cavalrymen supported by one so-called tankette, to destroy the intruders and "punish" the Soviet aggressors.* An even more powerful force, consisting of eight hundred infantrymen of Colonel Takemitsu Yamagata's Sixty-Fourth Regiment, accompanied by an artillery unit, was sent to back up Azuma, and to strike from the north boldly down the Halha. The aim was to encircle the Soviet forces that entered the disputed territory between Nomonhan and the Halha, and destroy them. Accordingly, Azuma thrust down the east bank of the Halha, near the confluence of the small Holsten River that ran into the Halha from the east. He would turn and form an anvil on which Yamagata, swinging westward from Nomonhan, could crush the enemy with a hammer blow. That was the plan, and, on paper, it looked like a powerful riposte against a small force of largely Mongolian horsemen.

Overhead, the ubiquitous Tsuji had taken to the air to watch the unfolding action from a light aircraft. Landing, he pointed proudly to a bullet hole in the fuselage. But he had seen nothing to suggest to him that the enemy was being reinforced. He certainly could not have known that, in Moscow, merely from reading newspaper accounts that were based on Manchukuoan press releases, the Soviet High Command had become alarmed and had ordered General Nikolai Feklenko to send his Fifty-Seventh Army Corps, currently stationed in the Mongolian capital, Ulaanbaatar (or Ulan Bator), toward the

*The Japanese tankette was a feeble toy, carrying one machine gun and armor so thin that an ordinary small-arms round could penetrate it.

Nomonhan battle zone. Unaware of any pending crisis, the unhappy Feklenko was caught out, away on a hunting trip. Knowing, however, what failure under the Soviet regime could mean, he returned to base at once and promptly dispatched a scratch force toward the battlefield. It included heavy artillery, tanks, and armored cars, which Komatsubara's forces did not have, and against which they lacked any sensible protection.

On the ground, Komatsubara's attack swiftly ran into trouble. His was a complex plan that required careful coordination, but radio communication between Azuma and Yamagata's units broke down immediately as a result of "chronic equipment failures."* By May 28, Azuma found himself caught in a trap, relentlessly bombarded by Soviet 122 mm heavy guns, and harried by tanks—and unaware that Yamagata had become bogged down. Without effective antitank weapons, Azuma's men improvised with suicidal tactics. Infantrymen equipped with fuel-filled bottles and a short fuse (what would later become known as Molotov cocktails) would clamber atop enemy tanks, lever open the hatch, and throw their bomb inside. The fuel would ignite murderously inside the tank—but the odds that the heroic assailant would survive were low, as he would likely be either engulfed in the tank's flames or shot down by machine gun fire from the accompanying Soviet tanks. These men were well named as "human bullet" (*nikuhaku*) antitank teams. Azuma's force, writes Coox, was soon in grave trouble, "short of water, lacking time to eat or to tend to their horses since the preceding night, pounded by artillery, low on ammunition, and plagued by sandstorms that clogged their weapons."

Particularly grievous was the state of the horses, caught out in the open by Russian shellfire. Azuma's staff urged him to try to break out of the enemy ring while there was still time. But no, his instructions

*The dangerous Japanese penchant for excessively complex plans would be seen again, this time at a more decisive level, at the Battle of Midway in 1942 (see part 4).

(as drafted by Tsuji) were to resist to the end, and to hold fast to the river junction they had been ordered to secure. In the early hours of May 29, Azuma's men drove off a Russian grenade-throwing attack, storming out of their shallow trenches "like fierce lions and tigers." But casualties were high, and both of Azuma's company commanders were killed. Exposed in the bitter cold of the Manchurian nights, his men (from subtropical Kyushu Island) suffered badly, only to be faced with the searing heat of a waterless day. By the afternoon of May 29, the Russians were closing in for the kill. Calling on the last nineteen survivors around him to join in a banzai charge, Azuma leaped out of his trench and was cut down immediately by Soviet machine gun fire. Virtually the whole of his two-hundred-man force was wiped out. Meanwhile, Colonel Yamagata's main force had fared little better, losing some 25 percent as casualties. (Yamagata himself survived, only to commit suicide in battle three months later.) According to some Japanese accounts, before the end of the affray, Tsuji turned up at Yamagata's command post, an unwelcome arrival, to berate the colonel for not relieving Azuma, and ordered him to go out and personally bury the bodies of Azuma's men.

Despite this expensive reversal to Japanese arms, which had cost some two thousand casualties, Tsuji's office mystically proclaimed "one victory and one defeat" from Kwantung general headquarters. Implicit in this "victory" was the heroic death in battle of Lieutenant Colonel Azuma. After it, according to Japanese official history, "remorse ate at the heart of General Komatsubara." There was even a suggestion made by the Kwantung Army's commander, General Kenkichi Ueda, that if a fresh division were sent in to redeem this failure, Komatsubara should follow the example of Marshal Nogi and commit seppuku. Nevertheless, a new offensive plan would soon be in preparation at Hsinking under the aegis of Tsuji. Meanwhile, there had been a fact-finding visit from Tokyo by Tsuji's counterpart, Colonel Masazumi Inada, the chief of the operations section at the Army General Staff

(AGS). His first question had been "Where's Nomonhan?" Then he had returned to base with the usual limp recommendation when confronted by the *gekokujo* movers and shakers at Hsinking: leave matters in the capable hands of the Kwantung Army. And this even though he himself had come away with a "gnawing suspicion" that Hsinking did not fully comprehend the seriousness of the international situation or the reasoning that underlay the General Staff's caution. To lend additional weight to this caution, Stalin's new foreign affairs commissar, Vyacheslav Molotov, addressing the Supreme Soviet on May 31, gave warning that "by virtue of our treaty with Mongolia, we shall defend its frontiers as energetically as our own . . . patience has its limits."

It was the following day that General Zhukov was summoned to the Kremlin by Voroshilov and told to prepare for his "long journey." Five days later he would arrive at Fifty-Seventh Corps' headquarters in eastern Mongolia at Tamsag, one hundred miles behind the front on the Halha River.

At this point, timings become all-important in relation to the much bigger game that was about to be played out at the other end of the world. On March 15, 1939, Hitler marched into Prague, thereby occupying the rump of Czechoslovakia, in direct breach of the Munich agreement with Prime Minister Neville Chamberlain. With the destruction of Czechoslovakia went any hopes that still existed for the maintenance of peace in Europe. Clearly Poland would be next on Hitler's list. On May 22, as Komatsubara was launching his first attack against the Soviets in Mongolia, Italy and Germany entered into the Pact of Steel. Signed by Hitler and Mussolini, the treaty between Germany and Italy, otherwise known as the Pact of Friendship and Alliance, pledged each party to support the other in case of war. Even Chamberlain now prepared for war.

All through that menacing summer of 1939, Britain and France scrambled desperately, if not indecorously, to form counteralliances in eastern Europe. Stalin was not impressed by the feeble support they could offer. Not surprisingly, given their history, the Poles would not permit Russian troops to enter their territory. Unlike most west European leaders, the Kremlin had studied, and had taken seriously, Hitler's rantings in *Mein Kampf.* Poland, Stalin recognized, would be only a bridge to Hitler's main ambition, the subjugation of Soviet Russia. The situation in Europe was a keg of gunpowder; meanwhile, Russia's back door in the Far East was being hammered on by an aggressive and expansionist Japan. Although theirs was the bite of a flea compared with what Hitler could do in the west, the Japanese would, as a matter of urgency, have to be crushed ruthlessly before Stalin could feel safe to pursue any diplomatic maneuvers in Europe, and certainly before Tokyo might be tempted to join the Pact of Steel. In no way could the Soviet Union face a war on two fronts on two continents. Stalin was contemplating a nonaggression treaty with Germany, which, in August, would come to be realized in the Molotov-Ribbentrop Pact. Thus it came to pass that a great deal would depend on this little "border incident" in the faraway wastes of Mongolia, all of which was invisible to the players in Europe at the time.

Oblivious to all this, and to the state of the Soviet forces in Mongolia, the Kwantung Army pursued its plans as if in a vacuum. Its intelligence remained abysmal. By way of avenging the destruction of the Azuma-Yamagata column, a devastating air raid was launched on June 27 on the forward Soviet air bases of Tamsag Bulak (or Bulag) and Bain Tumen deep in Mongolian territory. Newly arrived Russian planes were just forming up there. One hundred and twenty Japanese planes took part, an unprecedented number for that time. Needless to say, the bombers were accompanied by the ubiquitous Tsuji, whose hand was to be seen in the planning of the raid. Exultant, he counted twenty-five enemy planes destroyed, and nearly a hundred more shot down by Japanese fighters as they struggled to take off.

As with so many events in this war, the raid was carried out entirely on the initiative of the Kwantung Army. Deceitfully, it was launched only hours before orders forbidding any such aerial activity arrived from the AGS in Tokyo. When he heard of this latest escalation in the war, the emperor was furious and prohibited any further air attacks into Mongolia. Colonel Inada, generally regarded as a friend of the Kwantung clique, telephoned to excoriate his opposite number at Hsinking as a "damned idiot," accusing him of indiscipline and bad judgment. But Tsuji and the disciples of *gekokujo* turned a blind eye, telling themselves that they were, as always, executing the real wishes of the emperor, surrounded as he was by weak-kneed and corrupt advisers. An incensed Tsuji took the opportunity to demand autonomy for the Kwantung Army in "trivial" border matters. He later wrote that "tremendous combat results were achieved by carrying out dangerous operations at the risk of our lives. It is perfectly clear that we were carrying out an act of retaliation. What kind of General Staff ignores the psychology of the front lines and tramples on their feelings?"

He characterized the conflict between the Kwantung Army and the Tokyo General Staff as typical of the antagonism between combat officers and "desk jockeys." But this embracing of downright insubordination even toward the emperor seemed to indicate a militarist regime that was running off the rails.

Seen purely as a tactical achievement, however, the Tamsag Bulak raid was a tremendous coup for the Japanese, gaining for them a mastery of the air over the battlefield that they never lost. What a tragedy it was for the Western world that the defenders of Pearl Harbor, Malaya, and the Philippines two years later could not have known of the skill with which the Japanese Air Force could attack a ground target. Many US and British lives might have been saved, not to mention great warships like the *Repulse* and *Prince of Wales*. As an immediate consequence, Voroshilov's rage in Moscow exceeded that of the mikado's in Tokyo. The heads of the Soviet airmen held responsible would roll. The deputy commander of the Mongolian army was denounced as a Japanese agent and saboteur and shot. Meanwhile, as a priority the newly arrived Zhukov would be sent every reinforcement he called for—and, as would prove to be his wont in the Second World War, he called for a lot.

It was thus in a somewhat charged atmosphere that Tsuji and the staff at Hsinking planned a grandiose July offensive, their considerable self-confidence greatly boosted by the recent aerial success. With the usual deficiency of hard intelligence, that self-confidence replaced an ignorance of the realities of the battlefield. All across the Japanese lines there was, however, "ample evidence of the transcendent importance of fighting spirit."

Kwantung aerial reconnaissance had spotted large formations of Soviet road transport heading westward from the front, and deduced that this indicated an enemy withdrawal. They were, however, quite wrong. What they saw were in fact empty vehicles returning from the frontline formations having delivered their heavy cargoes of fuel, ammunition, and food under the cover of night. It was a massive shuttle system that had been organized by Zhukov's predecessor, General Grigory Shtern, a phenomenal achievement of the rickety Soviet system whereby more than 4,200 vehicles trucked in some 55,000 tons of matériel from the distant railhead at Borzya, over 400 miles of rough, dusty tracks. By comparison, the Kwantung Army could muster only 800 vehicles. This supply feat would completely deceive the Japanese as they prepared to attack. At the same time, Japanese intelligence chose to disregard reports of a large number of enemy tanks moving up.

Since the Azuma disaster, for which he felt a deep sense of personal responsibility, Komatsubara had been committed to a period of what the Japanese call "self-reflection." Nevertheless, as Coox points out, "almost none of the lessons from the fighting in May had been mastered by the time the next stage of the Nomonhan Incident broke out." In broad terms, Komatsubara's offensive at the beginning of July would be a more ambitious variant of the Azuma-Yamagata plan. It would comprise a bold pincer movement in which the main weight of the Twenty-Third Division, reinforced by a regiment from the Seventh Division, would cross the Halha and, sweeping deep into Mongolian territory, take the Soviet defenses from the rear. Simultaneously, the enemy would be attacked frontally by a powerful strike force commanded by Lieutenant General Masaomi Yasuoka and containing the Japanese army's only operational tank brigade. The principal objective would be the so-called Fui Heights, which in fact were no more than a "raised pancake" about a mile to one and a half miles across and no more than forty or fifty feet higher than the surrounding country. But, close to the confluence of the Holsten and the Halha, it was considered to be a commanding position in the forthcoming battle. The whole attacking force would total 15,000 men, 120 pieces of artillery, 70 tanks, and 180 aircraft. It certainly seemed a powerful body with which to correct a small boundary dispute.

On the night of July 2–3, while Zhukov was still lining up his forces for a massive riposte, the Japanese achieved a brilliant tactical success. A battalion of the Seventy-First Infantry Regiment rowed silently across the Halha River on a moonless night and made an unopposed landing on the west bank opposite the Fui Heights. Tsuji went along on the operation, which he had planned himself. The objective: to trap and destroy the enemy forces at the river junction. Conditions for the advancing Japanese troops were appalling; the lack of transport vehicles meant that they were forced to slog across sand baking under the fierce sun at over 100°F (38°C), fifteen hours a day, carrying eighty-pound packs. At night they froze on the shelterless steppe, tormented by the swarming mosquitoes. Soon there was an acute shortage of water. Men of one unit launched a crazed charge to capture a Russian machine gun simply so that they could drink its cooling water; others tried to assuage their thirst by licking the blood of their wounds. The plight of the horses was particularly pitiful. No lessons seem to have been learned since the disastrous Azuma operation in May.

Gradually, Tsuji's operation began to peter out. The daring raid across the river had managed to establish only one bridge, and this was so rickety as to make passage by tanks and heavy equipment impossible. So the Japanese infantry who had established the bridgehead found themselves without armored support. Russian artillery fire, inflicting heavy losses on the crews, effectively destroyed the Japanese bridging equipment. General Komatsubara, in a state of misery, decided to withdraw his tenuous bridgehead across the Halha. While the new offensive was still in the planning stage, back at Hsinking general headquarters, General Inada had remarked sadly that if he were Komatsubara "he would kill himself." The moment was approaching when the unfortunate Komatsubara would feel somewhat the same.

Meanwhile, on the other side of the lines, General Zhukov was taking over at his Mongolian headquarters. Furious at the success of the Japanese air raid on Tamsag Bulak, his rage had been compounded by the fact that the Japanese crossing of the Halha had caught him on the back foot, before he had been able to assemble his sledgehammer blow. Had it been backed up by greater force, Zhukov recognized late in the day, the move could have led to a wedge threatening his whole forward position. Knowing the fate of so many of his fellow generals, Zhukov may well have felt an uneasy twitch in the hair at the back of his neck. Meanwhile, he had damned his predecessor, Nikolai Feklenko, for being "badly organized, and [having] poor intelligence." The disorganization of the May battle, reported Zhukov, was the result of "poor tactics, ill-conceived battle management, and a failure to calculate and anticipate enemy manoeuvres." On arrival he had asked Feklenko tartly whether operations could be directed seventy-five miles behind the fronts, and it had soon become clear to him that the Fifty-Seventh

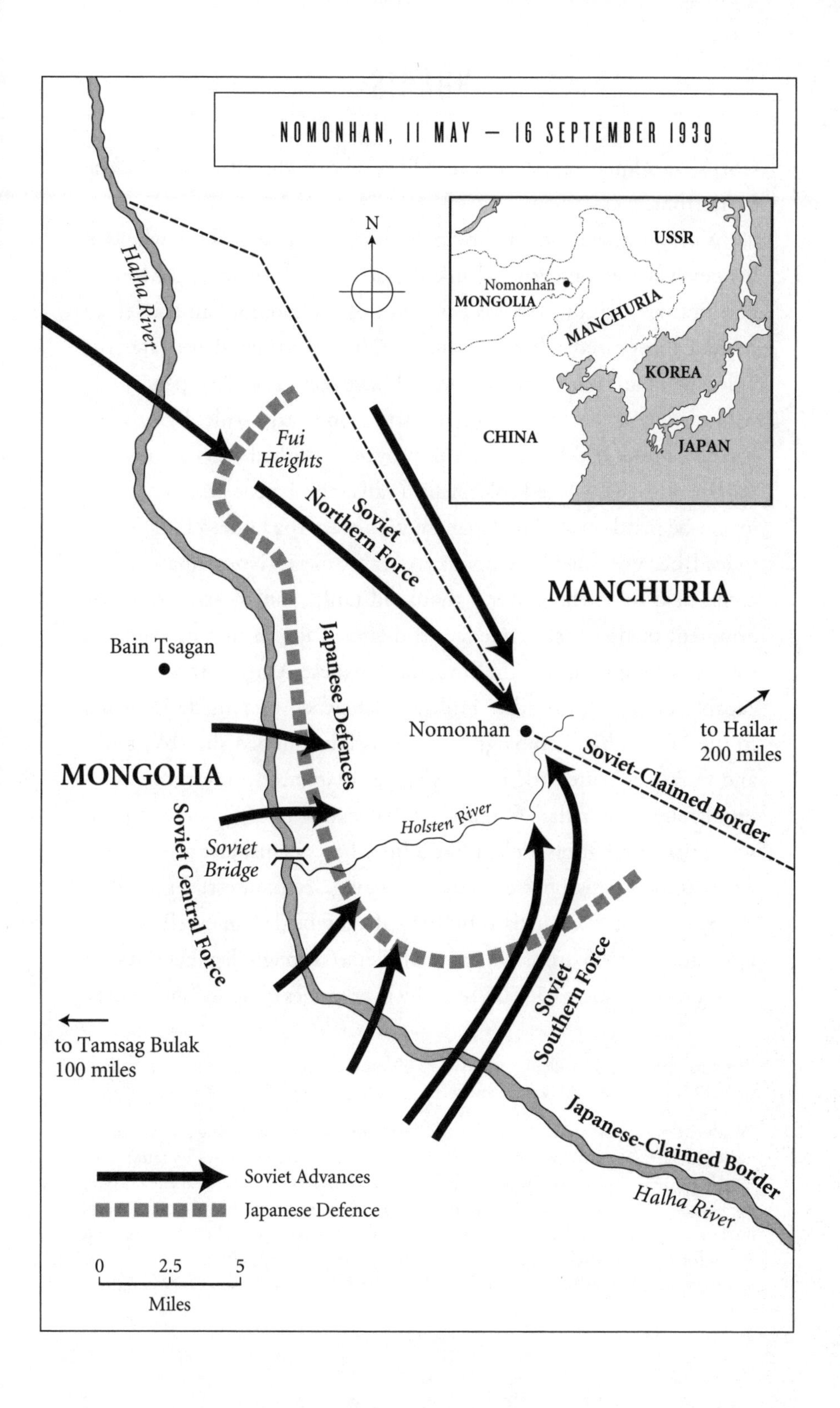
NOMONHAN, 11 MAY – 16 SEPTEMBER 1939
N
USSR
Nomonhan
MONGOLIA
MANCHURIA
KOREA
CHINA
JAPAN
Halha River
Fui Heights
Soviet Northern Force
MANCHURIA
Japanese Defences
Bain Tsagan
Nomonhan
to Hailar 200 miles
Soviet-Claimed Border
MONGOLIA
Soviet Central Force
Soviet Bridge
Holsten River
Soviet Southern Force
to Tamsag Bulak 100 miles
Japanese-Claimed Border
Halha River
Soviet Advances
Japanese Defence
0
2.5
5
Miles

Corps' headquarters staff knew little about the situation on the battlefront.[*]

Without hesitation Zhukov now threw in piecemeal the whole of his newly arrived Eleventh Tank Brigade and elements of a Mongolian cavalry division. On his first sight of the Nomonhan battlefield he had recognized what that mastermind of armored warfare, the fallen Mikhail Tukhachevsky, would have seen: that the open, sandy wilderness was ideal for armored warfare on a large scale. By July the Russian tanks had been much improved; they had external netting to deflect the Japanese Molotov cocktails, and hatches that could no longer be pried open. The Japanese *nikuhaku*, the human bullets, did indeed become suicide weapons. At the time of Nomonhan, Japan could field somewhere over a thousand tanks, but most were poorly armored, inadequately armed, and designed for an infantry-support role. This emphasis left the Imperial Japanese Army without a tank capable of taking on other tanks, a deficiency that made itself felt in the Nomonhan campaign.[†] The Russians dubbed the IJA's tanks and tankettes "cute little toys," whereas a Japanese infantryman gazing through binoculars found it "embarrassing to say," but "the Japanese tank-gun barrels looked like a little boy's penis, just protruding, whereas the Russian barrels were really long. 'No contest!' I thought."

By contrast, the Russians by 1939 already fielded an excellent antitank gun, the 45 mm, which could slice through Japanese armor, which was of a mere 15 to 17 mm thick, at ranges of up to 1,500 yards.

*Perhaps surprisingly, Feklenko survived to become an army commander in the disastrous 1941 battles in the Ukraine and went on to live through World War II altogether.

†With the priority of steel, which was constantly in short supply, being granted to the navy, the Japanese army was never able to rectify the shortcomings of its tanks. The warning of Nomonhan was slow to be recognized. After 1941, and with the entry of the United States into the conflict, priorities shifted to providing the weapons of naval warfare and defending the empire from the advancing Americans. Thus the tanks that Allied forces in the Pacific faced were primarily designs of the 1930s, such as the Type 97 medium and Type 95 light tanks, which had been Japan's mainstays at Nomonhan.

The Russians had also improved their main battle tank, the BT-7, one of the most mobile in the field by the standards of 1939, capable of a fast 40 mph (it could also run on wheels on road, by shedding its tracks, which was unusual, but of little use in Mongolia where there were no roads). Its revolutionary suspension had been stolen from an American designer, J. Walter Christie, and would subsequently be used by the Russians in the T-34, the most successful tank of the Second World War.* At only 14.5 tons, the BT-7 was still sufficiently armored to stand up to Japanese antitank weapons, and its 45 mm gun could knock out any of the "toy" Japanese tanks that came within its range. Nevertheless, one heroic Japanese battalion commander, Major Adachi, was watched by his men as he mounted a BT-7 with his sword drawn, only to be cut down by a tracer round fired from another tank.

In the torrid heat of a Mongolian July, General Komatsubara slowly lost the initiative, his offensive across the Halha being ground to a halt by Zhukov's improvised counterattacks. There now took place the first major tank-versus-tank battle in the history of war. It was hugely costly to both sides. The Japanese attacked with the whole of two tank regiments, which in fact comprised almost all the armor of the Kwantung Army. The Kwantung Army's standing regulations forbade the abandonment of a tank under fire, even if the machine was knocked out. Crewmen were ordered "to share the fate of their tank, saving their last bullet to commit suicide." But many Japanese infantrymen in the July battles claimed that they never saw a friendly tank in action. "Retreat and surrender were not merely dishonourable, they were literally impermissible. In some cases they could be capital offenses

*Instead of adopting the native Christie suspension, which gave tanks greater speed, mobility, and a lower silhouette, American tank designers in the Second World War went in for modified tractor suspension in their Grant and Sherman main battle tanks. The tanks were cheaper to produce—but they were not cheap when it came to the lives of crews. British tankers nicknamed the Sherman "the Ronson": "one flick and it lights."

punishable by death," comments the historian Stuart D. Goldman. It was a throwback to Japanese samurai concepts of martial honor. To add to the problems of the Japanese tank crews, the Russians also deployed a cunning "secret weapon": rolls of high-tensile piano wire that became entangled in the tracks of the lightweight tanks, bringing them to a standstill to be picked off by Russian guns. Nevertheless, the tank battle was also costly to the Russians. In the counterattacks, Commander Yakovlev of the Eleventh Armored Brigade, who personally led from the front, was reported as having died a "glorious death." He was subsequently made a Hero of the Soviet Union, and the tank in which he died still stands as a monument on the battlefield today.

The most damage was wreaked by Zhukov's heavy artillery, superior in both range and caliber to that of the Japanese. Early on in the battle, a heavy Soviet shell landed squarely on Komatsubara's Twenty-Third Division headquarters, killing his chief of staff, Colonel Ouchi, and inflicting terrible carnage among the tethered horses. It shook morale badly, even affecting Tsuji, who was in the headquarters at the time. Japanese 120 mm guns, with their feeble maximum range of 5,500 yards, could hit back only if deployed close to enemy lines, thus exposing them to deadly counterbattery fire. In contrast to the Japanese, who had expended two-thirds of their ammunition in the first two days of the offensive, the Russians—despite their immense supply problems across Mongolia—seemed to have an inexhaustible munitions stockpile. To some Japanese officers, writes Goldman, the battlefield "seemed like one vast Soviet firing range."

By the end of the second searing week of July, Komatsubara had lost his hard-won bridgehead across the Halha, though he maintained a tenuous foothold just east of it, provocatively still in territory claimed by the Soviets. The Holsten pincer attack had failed, yet the Japanese command remained typically upbeat. The Kwantung Army commander General Ueda sent Komatsubara a telegram congratulating him and praising the "decisive combat results." Komatsubara relayed the commendation to his subordinate commanders. Once more this was wishful thinking. Overall Japanese casualties had now risen to 15,500, and forty-five planes, all their tanks, and much of their artillery were destroyed.

Soviet losses too had been heavy, with more than 5,000 casualties by the end of July, but Zhukov now had a ready supply of fresh reinforcements flowing in across the Trans-Siberian Railway. In contrast, Komatsubara had virtually no reserves coming to support his battle-worn and greatly diminished units. All air attacks on Mongolian bases were now strictly forbidden by Tokyo—a serious handicap. The fault lay squarely with the Kwantung Army; since its commanders had lied so brazenly about the scale of the Nomonhan "border incident" in the first place, back in May, they could not now call upon Tokyo for fresh reserves.

Thus, as Komatsubara's July offensive ended, he stood with some 70,000 fatigued and underequipped troops facing a high-grade force of 100,000 under Zhukov—with do-or-die orders from Stalin to win at whatever cost, and to do so swiftly. By August, Zhukov could call on 200 aircraft, 800 armored vehicles, 45,000 men in three rifle divisions, and 200 heavy guns.

The hand of a ruthless commander had made itself felt the moment Zhukov arrived at his frontline headquarters. To Moscow he stigmatized his hapless predecessor, General Feklenko, as "an enemy of the people." Bulletins were posted warning his troops that, if they failed to carry out their orders, they would be brought before military tribunals and be severely punished. "I call upon you," urged Zhukov, "to show courage, manliness, audacity, braveness and heroism. Death to despicable cowards and traitors."

On July 13 he issued a decree announcing that two soldiers had been shot for cowardice. Six days later the Fifty-Seventh Corps was reorganized into the First Army Group and given operational independence, an arrangement designed to ensure that Zhukov could act without interference, except for instructions coming directly from the

General Staff in Moscow. The command picture was completed on July 31 when Zhukov was formally promoted from Komdiv (divisional commander) to Komkor (corps commander)—the equivalent of a four-star general in a Western army. It was a meteoric ascent for a young and relatively untried general. But Zhukov knew that his neck would be on the block in the event of failure, or even of partial success. What he did not know were the motives in high-level European realpolitik that lay behind Stalin's priorities. Zhukov was in the Far East, not merely to punish a minor border incursion, but also to inflict major damage on an enemy power such as Tokyo would not forget.

By their very nature the Soviets were expert in the art of *maskirovka*, or deception. The Kwantung Army, with its weakness in intelligence work, lent itself as the ideal partner. Accordingly, a handbook called *What the Soviet Soldier Must Know in Defense* was distributed to Zhukov's units and deliberately leaked to the Japanese. Russian troops were observed to be digging in vigorously on the center of the line, across the Halha and opposite Nomonhan, apparently building defensive works—*maskirovka*! All this was precisely the kind of information that Tsuji and the apostles of hubris at Hsinking found most readily digestible. From it, they deduced that the July fighting had in fact battered the Russians into some kind of submission. But, thanks notably to Sorge's precise intelligence, Zhukov had very different intentions. On August 20, he first struck at the center of the Japanese line, from behind the defensive works that his troops had been so laboriously and visibly digging. Japanese intelligence at once concluded that this was the focal point of a new Soviet effort. Obligingly, Komatsubara sent forces to hit back at this part of the line. Soon he was committed, locked in a bitter fight from which it became harder and harder to disengage.

Then Zhukov pounced, attacking from both wings of the battlefield, his huge armor groups backed strongly by aircraft in a double-envelopment maneuver. The defenders had been reinforced by elements of General Kunisaki's Seventh Division, but these were paltry compared with the forces Zhukov had in hand. The flanks of the unsuspecting Japanese were held by Manchurian troops, poorly equipped and ill trained, in effect at best halfhearted colonial forces. (Comparisons would be made three years later, at Stalingrad, where the unfortunate Romanians were left holding the vital German flanks where Zhukov, once again, would strike with decisive force.)

The Japanese order of battle and the strengths and locations of its units were clearly itemized for Zhukov by Sorge's intelligence. On the southern flank, Zhukov's deputy commander, Major General M. I. Potapov, with his armor sliced through the inferior Manchukuoan forces, who were caught quite unprepared, and then struck brutally into Komatsubara's unprotected rear areas. A similar blow smashed through the Manchukuoan formations that were supposedly protecting the northern flank. Heavy artillery barrages and aerial bombardment of unprecedented violence rained down on the defenders. Japanese resistance was again impeded by the ban imposed by Tokyo on any further air strikes on Mongolian territory.

There would be three phases at Nomonhan: first, the brutal frontal attack to pin down the Japanese forward defense line, along the eastern slopes of the Halha; second, the penetration of both flanks, the attackers wreaking havoc among Japanese artillery batteries and command centers; and third, the wider envelopment thrusts aimed at Nomonhan itself and the total destruction of an army unable (indeed, forbidden) to escape.

There were two factors that principally dictated the force of Zhukov's attack. As the Japanese troops shivered in their lightweight uniforms and poorly protected shelters, cold winds at night heralded the end of the campaigning season and the onset of winter. But much more pressing was Stalin's urgent need to have a clear-cut victory in the east as he concluded his immensely risky maneuvering with the archenemy, Hitler, in the west. Starting on August 20, the assault had to be clinched within the week.

Zhukov's "softening-up" preparations on August 20 hit the Japanese positions like a cyclone. His heavy artillery pounded the flimsily built shelters without letup for as long as ten hours, his men having an apparently bottomless supply of shells. Komatsubara's men had of course received no warning of the attack from Tsuji's staff. By midday, units were experiencing great difficulty in obtaining supplies of ammunition, food, and water. On the southern flank, Potapov, smashing through the Manchukuoan units holding it, bent back the line some ten miles. The Manchukuoans fled. Far to the north, the two Mongolian cavalry regiments supported by Soviet armor easily routed the opposing Manchukuoan cavalry flanking the key Fui Heights. In the Japanese cavalry that faced them, 80 percent of the horses were killed that first day; the rest stampeded or were injured. "Although the sight was grisly, the horse soldiers were relieved to a certain extent because there was neither water for the animals nor time to attend to them," writes Coox. The commander of a reconnaissance regiment, Lieutenant Colonel Ioki, had no water of his own but depended entirely on the division's trucks to haul in supplies daily. Now the vehicles could not get through. Against such an emergency, the Japanese had dug ten wells, each around fifteen feet deep, but from them only a trickle of water could be drawn with mess tins. Every well had been smashed by the bombardments. "The men used towels to wipe up night dew and then chewed on the cloth to wet their parched throats." The Russians now moved up guns of every caliber, including 152 mm, and blasted the strongpoints at short range, firing over open sights. Flamethrowers burned out bunkers and dugouts, followed by infantry who moved in with grenades and bayonets. At the Fui Heights, a desperate charge was attempted by the First Company of Major Ikuta's First Battalion (Twenty-Sixth Infantry Regiment), but it was abandoned after the loss of sixty-five of its eighty-five men. One badly wounded soldier wrote despondently: "To be killed is all that remains."

The attackers seemed to know the location and the fighting value of every single Japanese and Manchukuoan unit. How did they know?

Where could the crucial information possibly have come from? Who would have thought that the source was a fun-loving German journalist living high on the hog in Tokyo itself? It was not until 0900 on August 21 that Kwantung Army headquarters learned officially from the Twenty-Third Division about the attacks that had been launched more than twenty-four hours earlier.

By the end of the twenty-first, the inner prongs of the Soviet envelopment were gradually constricting the enemy, on the wings in particular. A Japanese staff officer reported that the Twenty-Third Division's defenses had already been "split widely." The following day, August 22, when the unfortunate General Komatsubara called upon a regimental commander (Colonel Sumi of the Twenty-Sixth Infantry Regiment) to mount a counterattack, Sumi responded in a manner seldom heard in a Japanese army unit: "No sir, it can't be done. My unit is a regiment in name only. I have only about two companies and a battalion commander with me." Later Sumi explained that "I felt that Komatsubara's plan was suitable for a picnic not combat." But he replied to the general, "I am thinking of fighting and dying where I am located. Please let me do so."

By the end of August 23, Potapov's armor had broken the whole Japanese defensive structure. Accounts by the outgunned and outmaneuvered defenders caught up in the merciless fighting make poignant reading. One of Ioki's reconnaissance companies, for instance, was reduced to four or five men. The walking wounded were ordered to accompany their comrades, but those who felt that they could not make it were given hand grenades, the implication being that they should kill themselves rather than be taken prisoner. A sickening incident took place when Russian tanks of the Ninth Armored Brigade, breaking through deep behind the enemy lines, overran a Japanese field hospital (though the Japanese claimed that it was liberally covered with red crosses), crushing the wounded on their stretchers.

On August 24, Tsuji happened to be present at the command post of Major General Morita commanding the Fourteenth Brigade of the newly arrived Seventh Division. There Soviet armor had overrun a quartermaster's dump, destroying valuable food supplies. "Barrels of pickled apricots were breached, drenching the soil with red juice that looked like blood," in plain sight of the hungry defenders. Sitting down to eat that night, Tsuji and Morita were surprised by an enemy air attack; dowsing the candle, they jumped into an antiaircraft trench. If hostile aircraft could attack by night as well as day, Tsuji feared, that would mean that the Russians had regained aerial supremacy. At that moment, a subaltern burst in and shouted that the Fui strongpoint had fallen, together with every officer and man. Tsuji reacted furiously, berating the wretched man: "What do you mean by 'annihilated'? You are alive, aren't you?"

When Japanese counterattacks did take place—and not until the fifth day of the Soviet onslaught—preparations were made on the spur of the moment, and planning and execution were generally slipshod. Tsuji flew back to Hsinking to report on the situation at the front. There he encountered disbelief, almost amounting to paralysis, as the headquarters staff took in the full extent of the disaster unfolding at Nomonhan. From the well-marked map taken from a dead Soviet officer, Tsuji was able to tell them that the size of the attacking Soviet forces apparently exceeded the Intelligence Section's previous estimates by a factor of two, and exceeded Japanese strength by a factor of four or five. On his way back to the front, his plane was shot down. Indestructible as ever, however, he managed to make his way to Major General Kobayashi's command post on foot, which placed him closer to the fighting than even he had anticipated. There one of Kobayashi's units (he commanded Komatsubara's main infantry resources) had charged into the enemy positions and been almost annihilated by Soviet tanks. Casualties were becoming insupportable for the Japanese. Frequently there were cases of a mere corporal become an acting company commander, and a first lieutenant taking over a whole battalion.

The list of senior officers killed or wounded by August 25 made somber reading: Colonel Sakai's Seventy-Second Regiment alone had lost 323 killed (including thirty-one officers) and 377 wounded (twenty-one officers, including Colonel Sakai himself). Kobayashi himself had been hit in the thigh by a tank shell fragment and fell to the ground, bleeding profusely. The troops ahead of him broke under fire. Fleeing "like a tidal wave," they never noticed or else ignored the fallen general, whom they trampled with their boots until his wound became gory and filthy.

Equally hard hit by the long-range Russian 152 mm guns were the Japanese artillery. Exasperated by the nonstop shelling, a commander of a heavy artillery regiment, Lieutenant Colonel Someya, decided to commit suicide after two of his battery commanders had been killed. Having destroyed his heavy guns, Someya made a valedictory speech exhorting his men to die "like beautifully falling cherry blossoms." Then he killed himself.

Komatsubara's men more or less gave up their counterattacks after August 26. Soviet infantry of the southern and northern forces had further shrunk the envelopment trap by that day. Japanese bastions were falling in ever more rapid succession, and reinforcements were halted by the Soviet air force as escape routes to the east were severed. More and more senior commanders were killed, several by their own swords. Still, fanatical resistance continued in pockets. "We will be annihilated," said a young second lieutenant. "But we are Japanese soldiers," was the regimental headquarter's automatic response. Accounts of suicidal bayonet charges, in keeping with Kwantung indoctrination, proliferated. One regimental commander shot his wounded men with a revolver, at their own request, "sending them to heaven."

By the twenty-eighth, the Soviet net was closing in on Komatsubara's divisional command post, now reduced to no more than "an open, conical dip hardly more than 150 metres in diameter and three metres in depth." Tsuji, on his fifth voluntary visit to divisional headquarters, found the situation almost untenable, the general helpless. He later wrote, "It was like a living hell" as he listened to the moaning of the wounded. For the sake of honor, Tsuji urged that the divisional commander should be saved. But Komatsubara decided not to try to break out. Meanwhile, in a nearby bunker that housed the commander of the Seventy-First Regiment, Lieutenant Colonel Higashi, the colonel pleaded with the severely wounded to "please kill themselves bravely when the enemy approached to close quarters." That evening Higashi's men saw him leap out of the trench, brandishing his sword and shouting defiantly as he charged a Soviet position: "I am Lieutenant-Colonel Higashi, forty-nine years old!" He was instantly cut down, his intestines ripped open. Yet his mind remained lucid, and he lay in the dugout continuing to give orders, telling his orderly to report to Komatsubara that he had conducted "an unskilful operation but I could not do better."

By now, Komatsubara's headquarters was experiencing a hell of its own. When a shower of hand grenades was thrown in at point-blank range, his chief of staff Colonel Okamoto had his knee torn to shreds. Under intense fire, surgeons successfully amputated his shattered leg with only a local anesthetic. "Okamoto uttered no sound during or after the operation." Although Okamoto had lost a lot of blood and said he felt cold, he never complained. Instead, locking his hands behind his head, he was reported to have gazed at the sky, remarking, "My, but the stars are really beautiful tonight."* Such was the innate stoicism of Japanese soldiers of the period.

By August 27, Zhukov felt confident enough to issue a decree to the First Army Group announcing that the Japanese forces on Mongolian territory at Khalkhin Gol had been surrounded and de-

*Remarkably, Okamoto survived, but he was murdered mysteriously in a hospital bed the following year.

stroyed. What remained was the mopping up of isolated groups of a thoroughly beaten force. Nomonhan was reoccupied, but Zhukov adhered strictly to his orders from Stalin not to occupy an inch of territory beyond the claimed antebellum border. This had been a battle only to destroy a military threat. Repeatedly the words appear in the Red Army's orders: "We must teach them a lesson." The result: a total defeat for the Japanese, though this would never be accepted by Tsuji and the Kwantung Army. For neither side was it a cheap lesson. Long afterward, vultures as big as men were observed hovering over the grisly field of battle. Casualty estimates differ widely. The Japanese officially reported 8,440 killed and 8,766 wounded, and the Soviets initially admitted 9,284 total casualties. It is likely that these figures published at the time were reduced for propaganda purposes. Some sources claim that the Japanese suffered 45,000 or more soldiers killed, compared with Soviet casualties of at least 17,000. In recent years, with the gradual opening of the Soviet archives in the 1990s, a more accurate assessment of Soviet casualties has emerged, citing 7,974 killed and 15,251 wounded. In a 2001 publication, the Soviet losses are given as 9,703 killed and missing. Given that Zhukov was never one to blanch at heavy losses, it seems that the combined Soviet and Mongolian casualties were considerably higher than admitted. On both sides many died in the hospital from disease.

Japanese casualties were particularly appalling in relation to the number of participants. Of the 15,975 officers and men of Komatsubara's Twenty-Third Division, 30 percent were killed in action, 34 percent wounded, 4 percent missing, and another 8 percent reported as sick—a staggering loss ratio of 76 percent.

There were also some three thousand Japanese prisoners of war taken. The figure may seem suspect, given the prevailing Japanese ethos. Mention has already been made of the spirit of "no surrender" imbued in the troops by Tsuji and other fanatics of the Kwantung Army. The precept ran that a soldier taken prisoner disgraced his emperor and his family and was "to be despised and shunned as a coward." The harsh code of Bushido pursued returning Japanese POWs even well after the war. Captivity was a matter of eternal shame that had to be concealed from everybody, including one's parents and kin. Postwar returnees were subjected to rigorous courts-martial, followed, if they were lucky enough to escape execution or lengthy prison sentences, by "spiritual education," during which the subject was forbidden to have any contact with the outside world—certainly not his own family. It was hardly surprising that many opted to remain and face an even more uncertain future in Stalin's Soviet Russia.*

Though the Battle of Nomonhan was effectively won by the end of August, sporadic fighting continued into September as the Kwantung Army continued to believe in some kind of miracle. Finally an armistice was concluded on September 16. From different motives, both Tokyo and Moscow were deeply reticent: Japan out of the shame of defeat, Stalin through the complex workings of his deeply suspicious mind, fearful of any factor that might upset his current tortuous dealings with Hitler. In the *New York Times* of October 4, the paper's well-informed correspondent Hugh Byas reported that the scale of Japanese losses "had not been expected." His report was headed TOKYO ADMITS DEFEAT BY SOVIET; CALLS MONGOL BATTLE "DISASTROUS."

This was not, however, the spin that events would receive at Hsinking. Nomonhan had been a setback, yes, but nothing more. The hotheads there, and in Tokyo, were already at work on new plans. Next it would be the navy's turn. For years, Nomonhan would remain the unknown war, and not just because it had been eclipsed immediately by the more momentous event of the Second World War breaking out in Europe just as the fighting on the bleak steppes of Mongolia came

*There would be a curious inversion of this in 1945, when many thousands of Soviet POWs liberated by the Western Allies after the collapse of Nazi Germany opted to remain in the West and had to be repatriated by force.

to an end. Unlike the war of 1905, there were no naval or military observers present, such as the meticulous Captain W. C. Pakenham, and certainly no war correspondents.

On August 28, as the battle came to an end at Nomonhan, Zhukov started a letter to his wife, which he completed three days later:

> Hello, My Darling Wife
>
> Greetings and all my affectionate kisses to you. . . . Since 20.8. I have been conducting a continuous battle. Today the destruction of the Japanese Samurai will be completed. . . .
>
> Today I received the report that I had been awarded the title of Hero of the Soviet Union. Obviously you will know about this already. Such praise by the government, the party, and Voroshilov obliges me to try even harder to fulfil my duty to the Motherland.
>
> Affectionately and lovingly all my kisses to you
> Until we meet again soon,
>
> Zhorzh

In terms of public awards and citations, most of the laurels for the victory were heaped upon General Grigory Shtern, Zhukov's predecessor and nominal superior.* But within the conclaves of the Kremlin, where it really mattered, Zhukov was the man of the hour, the rising star. He had won a brilliant victory, the Red Army's first since the Civil War, exacting a suitable revenge for Russia's humiliating defeat in the 1904–5 Russo-Japanese War. From Nomonhan, like Napo-

*Shtern was arrested in June 1941 during a new purge of the Red Army, possibly one with anti-Semitic undertones, and was executed four months later. Following Stalin's death he was posthumously rehabilitated.

leon after the battles of Arcole and Rivoli, the ruthless young general emerged with more than just a reputation. He had gained invaluable command experience in battle. At Nomonhan he had employed the techniques that later would become his hallmarks: the ability, like Napoleon, to "see only one thing" and then concentrate all on a point d'appui of weakness. Proved adept in the art of *maskirovka*, he also displayed great patience in building up superior strength and then striking with devastating force at his enemy's most vulnerable point, coordinating massed artillery, armor, and motorized infantry with tactical air support—and having no qualms about accepting casualties when necessary. His superb deployment of the double envelopment would stand him well in front of Moscow in 1941, and again at Stalingrad the following year. There his tactics would be a carbon copy of Nomonhan. It could be said that, if Waterloo had been won on the playing fields of Eton, then Stalingrad was won on the bleak steppes of Mongolia.

In Tokyo, the failure at Nomonhan brought down the government, leading Japan a further stretch down the road to the military, fascist government of General Hideki Tojo—and to the Pacific War. Little more than modest strictures were meted out to the personnel of the Kwantung Army headquarters for involving the country in so foolish and disastrous a war. Writing in 1950, only months after receiving amnesty for alleged war crimes, Masanobu Tsuji, newly promoted to colonel, offered the following explanation, which was neither an apology nor an excuse: "I have to admit that we at KwAHQ could have acted more wisely, if we and those at AGS had kept cool heads."

The major consequence that the defeat at Nomonhan had for Japanese strategy was to cancel once and for all any contemplation of the "go north" policy, a decision that would be promptly, and most usefully, relayed to Moscow by Richard Sorge. The Japanese war party, still bent on expansion and the acquisition of raw material, now directed its energies to the "go south" policy: the attack on Pearl

Harbor, Singapore, and the Philippines. Even if the Imperial Japanese Army had learned nothing from Nomonhan, what it had picked up there strategically by way of battle craft undoubtedly helped it in its initial successes in the subsequent Pacific War. Despite his disastrous interventions at Nomonhan, the hand of Tsuji would be seen often in campaigns like Malaya, where he played a leading role in devising the tactics of infiltration that would shatter the British defenses. Those early triumphs would also cause the Western Allies to overrate the IJA, which they may not have done had they been able to observe the Kwantung Army's reverses in the summer of 1939 at Nomonhan.

For Moscow, the timing of the victory at Nomonhan had immeasurable significance. By August 31, Zhukov's forces had completed the third phase of their offensive and had essentially won the undeclared war at Nomonhan. The next day, on September 1, the Germans invaded Poland. The Second World War had begun. Stalin had played his hand with consummate skill. Realizing in the summer of 1939 that he would get nothing out of any defensive alliance with France and Britain to stave off Germany, and full of suspicion (not unreasonably) that their intentions were to get Germany and Russia committed in war against each other, he performed the most amazing volte-face of the age. He entered into a nonaggression pact with Hitler. The timing was critical. Stalin knew that Hitler intended to move on Poland in early September. Yet he could not afford to take any strategic risk in Europe so long as he had a belligerent Japan gnawing away at his backside in the Far East. On August 18–19, Hitler was applying maximum pressure on Stalin to receive his foreign minister, Joachim von Ribbentrop, in Moscow without delay. Only on the twentieth was Zhukov ready to launch his sledgehammer attack at Nomonhan. On August 23, Ribbentrop arrived in Moscow to conclude the nonaggression agreement that would seal Poland's fate and unleash war in Europe, the Molotov-Ribbentrop Pact. That was the very day, the first day, that Zhukov could report to Moscow, bullishly, that he was about to complete the total destruction of the Japanese forces at Nomonhan. With some glee, Ribbentrop passed on to Hiroshi Oshima, the Japanese ambassador in Moscow, the news that the Russians had assured him that they had won a great victory at Nomonhan. Oshima retorted that that could not be true, for an IJA staff officer who had passed through the Nomonhan area one or two weeks earlier had reported to him that the situation was "very favourable" to the Japanese.

Nevertheless, despite his diplomatic finessing, Stalin was still facing huge risks. Although he knew through Sorge that the Japanese army was not prepared for an all-out war with the Soviet Union, on past form he could not be absolutely certain that the Kwantung Army would act rationally, or obey instructions from Tokyo. What would have been the consequences if Zhukov had failed? With a still-menacing Japan barking at his rear, Stalin would hardly have risked marching into eastern Poland. Consequently, in launching Operation Barbarossa in June 1941, Hitler would have started a crucial couple of hundred miles farther down the road toward Moscow. Moreover, when the Nazis reached Moscow, the invaluable reinforcements from Siberia that turned the tide in its defense would not have been available. They would still have been locked up in the defense of the Soviet Far East against Japan.

Stalin's handling of the armistice negotiations with the defeated Japanese was masterly. Aware of the importance of "face" in Tokyo, he ensured that the victorious Russians would demand no reparations and make no territorial claims. Zhukov's forces would not move one inch beyond the border at Nomonhan. It was sufficient that, badly mauled, the Kwantung Army had "learned a lesson" and would not again commit aggression against its Soviet, or Mongolian, neighbors. Memories of Kuropatkin's disastrous campaign in Manchuria were still kept alive in the Kremlin. Here, too, was perhaps a precedent that General MacArthur could well have studied before his triumphant rush to the Yalu River in 1950.

The consequences of Stalin's brutal play of realpolitik with Hitler in August 1939 are etched in history. Once again, unhappy Poland would be subjected to another division. As the Wehrmacht smashed in the western doors of Poland, the Russians would move in to occupy its eastern provinces. That was the secret deal as agreed to in Moscow: Stalin would occupy the three Baltic states. A few weeks later, he would launch a dismal war against Finland, all with the objective of gaining a defensive belt for Soviet Russia against its new ally, Nazi Germany. But it would also bring the Red Army face to face with Hitler's Wehrmacht, with no buffer state of Poland between them any longer.

Perhaps, in the realms of what-if, a great deal of pain and loss on all sides may have been saved if the details of the Battle of Nomonhan had been more widely known at the time. Had Hitler, for instance, been able to take on board the full potential of the Red Army as evinced by Zhukov's performance, he may have been warier of launching Barbarossa in the summer of 1941. Then again, an awareness of the Japanese capacity for fanatical suicidal bravery at Nomonhan, as well as the Tojo regime's dedication to aggressive war, may have helped the United States in its conduct of the Pacific War. It may even have persuaded Roosevelt to place Pearl Harbor on something more of an urgent war footing. But these are lessons that, unseen, could never have been learned.

PART THREE

Moscow, 1941

CHAPTER 10

General Summer

IN THE COURSE of 1941, hubris would lead Hitler into committing in quick succession the three greatest mistakes of his career, mistakes of historic and fatal proportions. First was Operation Barbarossa itself, the decision to invade the vast landmass of Russia; second was his tardy commitment to an all-out attack to capture Moscow—a mistake his predecessor Napoleon had made a century and a half earlier; third, and the most ruinous of all, was his declaration of war on the United States on December 11, a mere week after his offensive on Moscow had finally ground to a halt. These were red-letter decisions of terrifying consequence.

It was, however, well before the first Panzers rumbled into Soviet space that Hitler's genius for war began to let him down, replaced by the hubris that would eventually destroy him and his hideous Third Reich. His strategic planning was grossly flawed. In the first place, his Barbarossa timetable had been disrupted by a quite unnecessary sideshow when he moved to crush Yugoslavia and clear the British out of Greece. That meant that, instead of starting his campaign in May, as soon as the quagmires of the spring *rasputitsa* (the twice-yearly mud season) had dried out and the short campaigning season had begun, Hitler delayed it until the last week in June. By a curious coincidence of history, he and Napoleon in 1812 set off within a day of each other. Napoleon delayed for the much more valid reason that he had to wait until the grasses of the steppe had grown sufficiently to fodder his horses. As it happened, it was not "General Winter" that defeated both Napoleon and Hitler, but "General Summer." In the short but intense summer heat of 1812, the horses of Napoleon's cavalry, as vital to his operations as the Panzers were to Hitler, perished from thirst and fatigue; Hitler's Panzers would arrive before Moscow worn out and disabled, hors de combat from the dust of summer before the first snows even fell.

On the eve of Barbarossa, there was a fundamental disagreement between Hitler and his senior generals over strategic priorities. The Führer's instructions were "Leningrad first, the Donetsk Basin [the great industrial area in eastern Ukraine] second, Moscow third." He had no doubts that his invincible Wehrmacht could march on all three objectives at once. To him, Moscow was of "no great importance"; instead, he saw victory achieved through the destruction of the Red Army west of the capital. In the first crucial weeks and months of the campaign, Hitler would have his way. Initially, as in France, the campaign went superbly well. On the very first morning, Stalin's unprepared air force lost over 1,200 aircraft, most of them on the ground, lined up wingtip to wingtip (many of the planes newly delivered).*

Hitler had been encouraged by the humiliation inflicted by the Finns on Soviet forces in the Winter War of 1939–40—a war triggered by Stalin's desire for a defensive buffer. Employing the crude, post-1918 tactics of infantry attacks uncoordinated with armor, such as Zhukov had discarded at Nomonhan, the Red Army in Finland suffered horrendous casualties at the hands of a small but doughty nation. According to Nikita Khrushchev, who was then head of the Communist Party in Ukraine, the Soviet leaders had believed that "all we had to do was

*In fact, Soviet losses were probably far higher. According to a postwar Russian historian, Viktor Kulikov, some 3,922 Soviet aircraft were destroyed over the first three days, 2,000 of them on that first day.

to raise our voice a little bit, and the Finns would obey. If that didn't work, we could fire one shot and the Finns would put up their hands and surrender." It was an illusion that was swiftly shattered when the Finns fought back with a skill that the Russians could not match. Despite the assurances of Kliment "Bag of Shit" Voroshilov, in command of the campaign, that "all is well, all is in order, all is ready," in 105 days of savage fighting the Soviets suffered 200,000 casualties, including 50,000 dead, out of a total of 460,000 men. Over 3,350 artillery pieces and about 3,000 tanks, a hundred times as many as the Finns had, supported by 1,300 aircraft, were deployed on the Karelian Isthmus alone. It was a miserable war for the Soviet soldier, ill equipped by comparison with his adversaries. As the historian William R. Trotter has written, he "had no choice. If he refused to fight, he would be shot. If he tried to sneak through the forest, he would freeze to death. And surrender was no option for him; Soviet propaganda had told him how the Finns would torture prisoners to death."

The following June brought Hitler's stunning blitzkrieg in France: the most powerful army in the west destroyed in six weeks at a cost in German lives not much higher than Zhukov's losses at that one small confrontation in Mongolia. Stalin was astounded and appalled, almost taking the French defeat as a personal affront. To Khrushchev he exclaimed, "How could they allow Hitler to defeat them, to crush them?" It upset all his calculations, which were predicated on a repeat of the costly and long, drawn-out trench war stalemate of 1914–18. Stalin had hoped the fighting would leave Germany so debilitated that it would be years before it could risk unleashing a war on the Soviet Union.

On June 2, 1940, as the French army was disintegrating and Churchill was pulling the British Expeditionary Force out of Dunkirk, Zhukov had his first meeting with Stalin. One can assume that Stalin was content to see the victor of Nomonhan back in western Russia. Three days later, Zhukov, still in his early forties, was promoted to the rank of full general.

There was one far from negligible reason why the Russians would be so badly handicapped in the first days of the fighting. When the Soviet forces had moved into the buffer zone of eastern Poland in 1939, Stalin had given priority to the grisly myrmidons of Lavrenty Beria's NKVD. Among a host of brutal arrests and executions of Poles was the massacre of some twenty thousand Polish army officers at Katyn in one of the worst atrocities of the war. It is hardly surprising that as a consequence the Soviet army was treated in Poland as an enemy force imposed upon a resentful and hostile population, rather than as an ally come to protect the Poles from a German invasion.

Especially grave was Stalin's own refusal to recognize the imminence of catastrophe. From the early months of 1941, warnings of Hitler's intentions came in by the score. They came from trusted Soviet agents (spies) like Anthony Blunt in London and Sorge in Tokyo; from Churchill, and from Roosevelt. The warlord, immured with his yes-men in the Kremlin, refused to believe any of them. On March 5, 1941, among many such warnings, Sorge sent to Moscow a microfilm of German documents indicating a German attack in mid-June. It received an icy comment: "We doubt the veracity of your information." His report was then banished to the limbo of "on file." For whatever reason, Stalin seems to have become personally disenchanted with his ace spy, and he was reported to have damned him, just before Barbarossa, as "this bastard who's set up factories and brothels in Japan and even deigned to report the date of the German attack as June 22. Are you suggesting I should believe him too?"

Under mounting pressure in Tokyo, Sorge was infuriated by the Kremlin's disregard for his information. "Why didn't Stalin react?" he asked, baffled. It strained his nerves at an increasingly difficult time, as the Japanese secret police were closing in on him. He took to drinking ever more heavily.

As late as June 15, when Zhukov and Marshal Semyon Timoshenko (who had replaced Voroshilov as commissar for defense) came to him begging for reinforcements on the frontier, Stalin waved aside their fears, declaring, "We have a non-aggression pact with Germany. Germany is up to her ears with the war in the west and I am certain that Hitler will not risk creating a second front by attacking the Soviet Union. Hitler is not such an idiot and understands that the Soviet Union is not Poland, nor France and not even England."

Three days later, on June 18, the two generals tried once again to persuade Stalin and the Politburo to put the army on full alert. The meeting lasted for three hours. Stalin became increasingly irritable, always a warning sign. He accused Zhukov of warmongering and "became so abusive that Zhukov fell silent." Stalin repeated: "But you have to understand that Germany on her own will never fight Russia. You must understand this."

Egged on by the murderous toady Beria, Stalin managed to persuade himself that all this intelligence was simply a plant, designed to set the Germans on the Russians. He would trust Hitler rather than Churchill, persuaded that, on past form, the Western democracies were more deeply committed ideologically to the destruction of the Soviet state than was Nazi Germany. In short, he refused to believe that a greater villain than Joseph Vissarionovich Stalin could possibly exist. On the evening of June 18, a German deserter came over with irrefutable intelligence on the date of Barbarossa. He was shot for his pains. All those last days, Soviet forward troops could hear the sound of the German Panzer mechanics tuning their tank engines. Yet, right to the very end, to the fateful June 22—and even beyond—Stalin would continue to insist that no provocation should be offered to the Germans. At the very last moment, Stalin decided to call up eight hundred thousand reservists—a process to be completed by July 10. It never was; it was too late.

By July 10, the Panzers were well inside the Soviet Union. Attacking all along the longest continuous front ever known, which stretched for 1,800 miles, from the Baltic to the Black Sea, the most fearsome attacking force in the history of war swept eastward. It totaled 3.8 million men (including some 500,000 from Hitler's eastern European allies such as Romania and Hungary), with 4,300 tanks, 4,389 aircraft, and 7,200 artillery pieces supported by 600,000 motor vehicles and 750,000 horses.

The land force comprised 148 German divisions, of which nineteen were Panzer and twelve motorized infantry. They were faced (on paper) by some 5.5 million Soviet men under arms, but just under 3 million of that number could be counted as frontline troops. The Soviets had around 20,000 tanks (many of questionable readiness for battle) and 35,000 aircraft—though only 11,000 were actually serviceable on June 22. The German phalanx was led by battle-hardened commanders such as Heinz Guderian, the progenitor of the *Panzerkorps*, who had romped through France the previous summer and more recently had crushed Yugoslavia. The Germans' self-confidence was high, and they regarded their forces as more or less invincible and the foe contemptuously as "a colossus with feet of clay." They estimated that it would take six to eight weeks to destroy the Soviet Union. Nevertheless, only a genius or a madman could have attempted such an invasion; in a short space of time, Hitler would prove to be more of one than the other.

General Dmitry Pavlov, the commander of the key western front, was enjoying an evening at the Minsk Officers' Club, together with his chief of staff, Vladimir Klimovskikh, and the district deputy commander, Lieutenant General V. I. Boldin. A popular comedy, *The Wedding at Malinovka*, was playing to a full house. The evening's pleasures were interrupted by Colonel Blokhin, the head of intelligence in the Western Special Military District, who reported to Pavlov that "the frontier was in a state of alarm." Despite the warnings, not a single order had so far been issued to either the Red Army or the navy.

On hearing the news of the attack, Zhukov rang Stalin early in the morning, to be told by an aide, "Comrade Stalin is sleeping."

"Wake him up immediately. The Germans are bombing our cities."

Three minutes later Stalin was on the telephone. Zhukov reported the situation, and was answered by silence. "Did you understand what I said?" Zhukov asked. Still there was silence.

Stalin still couldn't believe what he was hearing. Then, when told of the destruction of his air force, he remarked, "That's a monstrous crime, those guilty of it should pay with their heads."

Beria was ordered to investigate. Molotov wrung his hands and exclaimed, "What have we done to deserve this?"

If anybody should have "paid with his head" for the initial Soviet disasters, it should surely have been Stalin himself. In a democracy, he would no doubt have fallen, but such was the regime of terror he had created around himself that he was untouchable. For several days he retreated into his shell in his dacha, leaving Molotov to broadcast the terrible news to the people of the Soviet Union.

> Citizens and Citizenesses of the Soviet Union! Today, at four o'clock in the morning, without addressing any grievances to the Soviet Union, without declaration of war, German forces fell on our country, attacked our frontiers in many places and bombed our cities . . . an act of treachery unprecedented in the history of civilized nations. . . . The Red Army and the whole nation will wage a victorious Patriotic War for our beloved country, for honour, for liberty. . . . Our cause is just. The enemy will be beaten. Victory will be ours . . .

Later that same morning in Berlin, Hitler proclaimed, "Before three months have passed, we shall witness a collapse of Russia, the like of which has never been seen in history." The omnipotent Red Tsar, regarded as the unshakable "man of steel," seemed quite unable to focus. After a while, Stalin emerged from this semiparalysis, shaken but able to resume his viselike control as before. It might well be said that, with Stalin in those parlous days, the war was all but lost; yet without him it almost certainly would have been.

Only on July 3 would Stalin broadcast to the nation. That same day, Colonel General Franz Halder, Hitler's chief of staff of the army, noted in his diary that it was "probably no understatement to say that the Russian campaign has been won in the space of two weeks." Certainly the Panzers, having swiftly broken through the feeble, ill-organized crust of the Soviet border forces, had been slicing their way into Russia at a speed that made the previous year's French campaign look almost pedestrian. In May 1940, they had swept through to the Channel, two hundred miles in ten breathtaking days; in the first days of Barbarossa they would frequently achieve the same advance in a couple of days. A German veteran of the campaign, Axel von dem Bussche, who as a young captain in an elite infantry regiment fought his way to the outskirts of Leningrad and then wintered in Tsar Paul's Palace of Pavlovsk, told me many years ago how that first summer he "never saw a dead German, nor a T-34 tank."

Within twenty-four hours of the beginning of Barbarossa, the Germans had rent a great hole eighty miles wide in the Soviet defenses. By June 29, after one week's fighting, Minsk, the capital of Byelorussia (Belarus), one-third of the way to Moscow, had been taken; by July 10, the Soviet western front alone had lost 4,799 tanks, 9,477 guns, 1,777 combat aircraft, and 341,000 soldiers. In less than three weeks Field Marshal Fedor von Bock's Army Group Center had advanced nearly four hundred miles along the road to Moscow. Two hundred out of 340 military supply dumps fell into German hands in the first month. Writes Rodric Braithwaite, "Russian soldiers surrendered, in huge numbers, in formed units, without firing a shot." On July 16, Smolensk, on Napoleon's route to Moscow, and little more than two hundred miles distant from the capital, fell, while the beginning of August brought the Panzers to the gates of Leningrad in the north. Kiev, the capital of the Ukraine on the great River Dnieper, was reached, bypassed and encircled by September 16. Against the advice of his generals, Stalin ordered that Kiev be held, regardless of cost—another disastrous decision. A savage ten-day battle ensued in which the Soviets lost the staggering total of 450,000 men and nearly 4,000 artillery pieces from forty-three divisions engaged. Kiev fell, nevertheless.

The Panzer thrusts deep into Russia would surround and cut off immense bodies of ill-trained and poorly equipped Soviet troops. The technique was known as *Kesselschlacht*, or "cauldron battle." Once the cauldron was sealed, the German infantry would move in to mop up. Unimaginable numbers of prisoners were taken. Over four hundred thousand were taken on the frontiers, another six hundred thousand at Kiev: altogether, a total of over three million by the end of 1941. After Kiev, the Red Army defenders no longer outnumbered the Germans, and there were no more immediately available trained reserves. To defend Moscow, Stalin could field only eight hundred thousand men in eighty-three divisions, but no more than twenty-five divisions were fully effective.

The fate of the huge numbers of Russian POWs, who were not granted the protection stipulated under the Geneva Conventions, was appalling. Fewer than 40 percent survived, and many of them, on their return home, were condemned to the Gulag camps for having surrendered. It was Nazi policy to starve the prisoners to death as part of the strategy set out in Reichsmarschall Hermann Göring's infamous "Green Folder," whereby the entire urban population of Russia was to be "reduced," to generate an agricultural surplus to feed Germany and to create the promised *Lebensraum*, or living space, for German émigrés. Close on the heels of the invading army followed Heinrich Himmler's murderous *Einsatz* squads, who with equal efficiency rounded up Jews, Gypsies, and commissars to be dispatched in mass executions. It was all part of Hitler's ideological strategy against a whole people whom he considered to be *Untermenschen* and who were to be totally eradicated rather than simply defeated in battle.

Here was one more fundamental error that would lead to Germany's undoing. In many areas of the western Soviet Union, and especially the occupied Baltic states, the Germans were welcomed on June

22 as bringing freedom from Stalinist oppression. In the Ukraine particularly, where millions had starved under the communist "planning" of the 1920s and 1930s, the Wehrmacht was greeted with the traditional offerings of "salt and bread." But views soon changed as news of the excesses of the *Einsatzgruppen* and the fate of the POWs crept out. Red Army units caught up in a *Kesselschlacht* now fought on to the death or joined up to form partisan bands that wreaked terrible havoc behind the German lines. What opportunities Hitler threw away with his policy of brutality would be shown later in the formation of a whole army of Soviet renegades led by a general, Andrey Vlasov, who fought heroically until taken prisoner in July 1941.

Through August and into September, the massive Wehrmacht onslaught ground on through the relentlessly scorching plains, where there was seldom any cover for man or beast. Yet it was also grinding down the attackers. The demands made on men, horses, and machinery were gigantic, and unprecedented. Russian counterattacks were ceaseless, and there were repeated strikes far to the rear by increasingly well-organized groups of Soviet partisans. Hitler's war machine, after all, was an instrument designed for short, sharp blitzkriegs, not for a prolonged offensive over thousands of miles of hostile enemy territory. Commanders at the front were now aware that the machine was running down, just as the Russian dust destroyed the bearings and tracks on the Mark IV and Mark III tanks that had waltzed across France the previous year. Their concerns were relayed back to the High Command in Germany. Halder had recorded in his diary on July 3 that the campaign was already won, but by the beginning of August he had changed his tune and was admitting the truth more solemnly.

> We have underestimated the Russian colossus. . . . [Soviet divisions] are not armed and equipped according to our standards, and their tactical leadership is often poor. But there they are, and

> if we smash a dozen of them the Russians simply put up another dozen. The time factor favours them, as they are near their own resources, while we are moving farther and farther away from ours. And so our troops sprawled over an immense front line, without any depth, are subjected to the incessant attacks of the enemy.

The German rate of advance fell from twenty miles per day to five miles. None of these gloomy, almost defeatist thoughts would be allowed to trickle through to the Führer, guarded as he was by toadies like the "lackey" General Wilhelm Keitel of the Armed Forces Supreme Command. Allegedly treasonable reports, such as that "some Panzer Divisions have only ten tanks left," would not reach him.

But the reality was that, by September, none of Hitler's three ambitious objectives in Barbarossa had been achieved. Leningrad had been reached but not captured—after a brutal and bloody intervention by Zhukov, sent in by a desperate Stalin, the city remained in Soviet hands. So Hitler had to resort to siege warfare in an attempt to starve the great city to death.* Meanwhile, icy blasts of wind blowing off Lake Ladoga to the northeast were already bringing intimations that the dread Russian winter was not far away.

At the other end of the colossal front, Hitler had captured half of the Ukraine, including the capital, Kiev. But the rich eastern half, with industrial plums like Kharkov and the Donetsk Basin, remained out of his grasp. In the crucial center, the road to Moscow lay barred. It would in fact take Hitler nearly three months longer to reach the city's outskirts than it had taken Napoleon. For, despite their deserved reputation for lightning armored warfare, the Germans were almost as dependent as Napoleon on horses, and on the stamina of their men on the march. It would turn out to be their greatest weakness.

---

*An estimated nine hundred thousand civilians died during the nine-hundred-day siege, but Hitler's forces never entered the city.

At the end of August, the Soviet High Command, the Stavka, ordered Zhukov to command the Twenty-Fourth Army in a spoiling offensive at Yelnya, fifty miles southeast of Smolensk, to pinch out a German salient that was threatening the capital from the southeast. Under the threat of encirclement, the Germans started retreating, and on September 6 Yelnya was retaken. It was the most substantial reverse that the Wehrmacht had suffered up to that date and the first successful Soviet offensive operation since Barbarossa began. As Rodric Braithwaite stresses, Yelnya reinforced Zhukov's reputation with Stalin as a ruthless commander, "the right man for a rough affair," in Molotov's phrase.

It was, however, a typical Zhukov battle, fought without regard to casualties—a pyrrhic victory for the Red Army. The overall Soviet losses estimated at 31,853, the battle seriously drained the forces left to cover Moscow after all the disasters of the summer, resulting in another huge *Kesselschlacht*, trapping a further half-million men. Meanwhile, however, another factor had emerged that would tilt the war in favor of the defenders: the T-34 main battle tank, which now began to make its appearance in mass formations on the battlefield.

Developed only the previous year from the light BT-26 tanks Zhukov had deployed at Nomonhan, the new T-34 was unquestionably not just a battle-winner but a war-winner, and it would generally be regarded by all sides as the best all-around tank of the Second World War. After many upgrades and improvements, its basic design would go all the way through to 1945 and beyond, when it would appear in the Korean War and even in various conflicts in the Middle East. Based on the Christie suspension, which gave it excellent cross-country performance though it weighed thirty tons, it could move at over 30 mph. A low profile of eight feet made it an elusive target. Carrying a high-velocity 76.2 mm gun, it was the most powerful anti-tank weapon on the field. No German tank of 1941–42 vintage could stand up to it, and the T-34's thick sloping armor made it hard to knock out. The T-34 ran on tracks substantially wider than its German rivals, which meant that it could operate on snow, ice, and the mud of the autumn *rasputitsa* where Guderian's Panzers failed. It was a fundamentally simple, indeed crude, design, which allowed it to be produced in large quantities. If the early T-34s had a flaw, it was their very simplicity: only the unit commander's tank carried a radio, and the cramped turret had room for only two, so the commander had to double up as gun-loader—an almost impossible duplication of tasks.

Having trained in spacious Sherman tanks during the Second World War, I can vouchsafe that no Western tankers would have put up with the crudeness and angular discomfort of the T-34. Nevertheless, facing the Barbarossa onslaught with only handfuls of T-34s at the outset, Stalin could field nearly three thousand by the end of 1941—to the shock and distress of the attacking Germans.

At last, disturbed by these factors, together with the waning of the brief campaigning season in Russia, Hitler began to change his mind. Working on him most forcefully had been Field Marshal Fedor von Bock, the commander of the key Army Group Center, the general closest to Hitler's Nazi ideology. A leader who lectured his soldiers about the honor of dying for the German Fatherland and nicknamed "Der Sterber" (literally "the Killer"), Bock had led Army Group B during the French campaign in 1940. Along with almost all the German frontline commanders in Operation Barbarossa, he had constantly, and vehemently, urged Hitler to concentrate on an all-out drive toward Moscow. Besides the psychological importance of capturing the enemy's capital, these commanders had pointed out that Moscow was a major center of arms production and the center of the Soviet communications and transportation systems. Also, intelligence reports indicated that the bulk of the Red Army was deployed around Moscow for an all-out defense of the capital. Thus a coordinated offensive toward Moscow could force the Soviet Army into a final, decisive confrontation. Hitler, however, had remained immovable, ordering his Panzer specialist, Guderian, to send Army Group Center's tanks to the north and south, temporarily halting the drive to Moscow.

Accordingly, on September 9, Bock was instructed to prepare an operational order for an all-out drive to capture Moscow. Under the code name Operation Typhoon, the assault on Moscow was to begin no later than September 30. Bock's Army Group Center would be reinforced and replenished with men and vehicles. It would be composed of three infantry armies (the Second, Fourth, and Ninth) and three tank armies, representing the cream of the most powerful remaining elements of the Wehrmacht forces in Russia.

On September 16, Bock issued his operational directive for Operation Typhoon. The three Panzer armies were to spearhead the attack. The Third (under Hermann Hoth), which had been withdrawn from the drive on Leningrad, would lead in the north. The Fourth (under Field Marshal Günther von Kluge) would lead the drive in the center, and Guderian's Second Panzer Army would thrust up from the south after its victory in the Ukraine. Hitler considered that the besieged and starving Leningrad could be allowed to wither on the vine, while in the Ukraine he would halt any move to thrust further east, into the rich area of Donetsk or toward the Caucasus. That could be deferred, if necessary, to the spring of 1942, once Moscow was in German hands. His strategy was to carry out another irresistible *Kesselschlacht* to send outflanking forces to the north and south of Moscow, then close in to cut the city off from the rear. But did he now have the wherewithal? Frontline commanders like Bock and Guderian thought not, but kept their doubts to themselves. No one knew better than Guderian, who had nursed the Panzer weapon from infancy, just how exhausted his armored spearheads were. And what about the dreaded Russian winter, which was now just around the corner?

Bock had assembled over a million men: three armies, three Panzer armies, and seventy-eight divisions, with 1,000 tanks and 14,000 guns, backed by 600 planes, representing the best, the most elite, that

Admiral Togo leading the action at the Battle of Tsushima in his flagship, *Mikasa*.

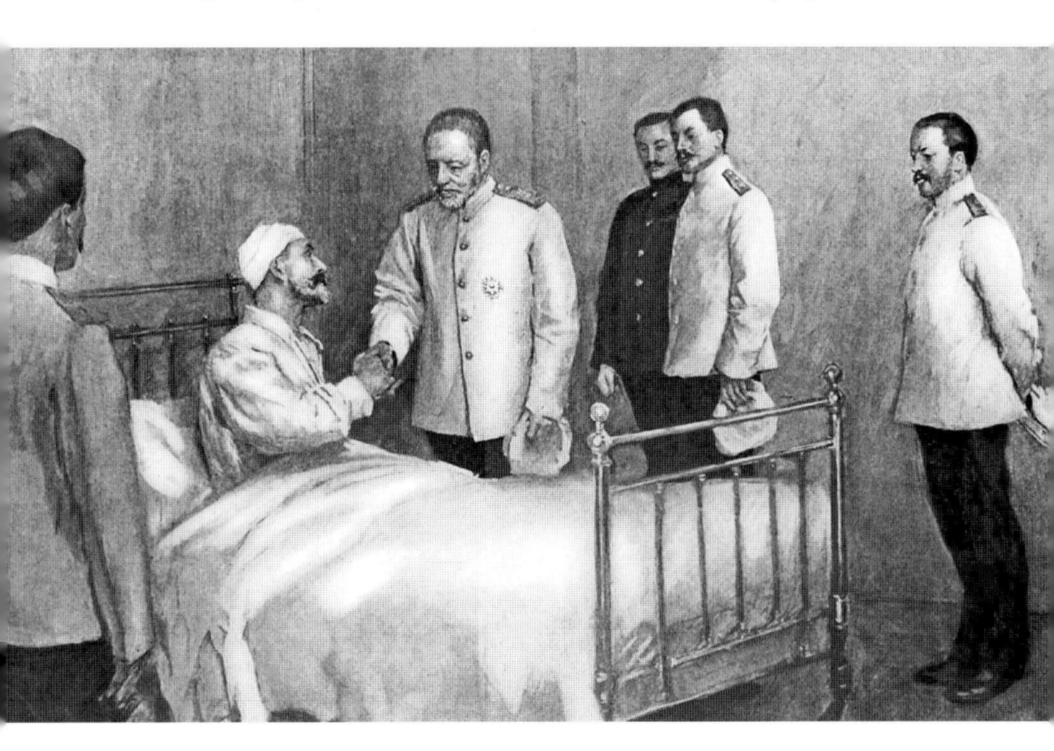

Admiral Togo visiting the defeated Admiral Rozhestvensky in the hospital after the Battle of Tsushima, 1905.

ПОЧТА СССР 1965
1895 — 1944
ГЕРОЙ
СОВЕТСКОГО
СОЮЗА
4
РИХАРД ЗОРГЕ

Marshal Georgy Zhukov presenting the Soviet Order of Victory to Field Marshal Bernard Law Montgomery, June 5, 1945.

Two Wehrmacht soldiers on guard west of Moscow, December 1941.

Opposite page, clockwise from top left:

Georgy Zhukov *(center)* among the Soviet generals at Nomonhan.

A Soviet postage stamp featuring a portrait of Richard Sorge, 1965.

Soviet soldiers inspecting a captured Japanese tankette—a Type 95 Ha-Go—at Nomonhan.

The God of Operations: Colonel Masanobu Tsuji.

A hastily assembled force of Moscow women gouge a huge tank trap to halt the Panzers.

The road-oriented Wehrmacht force of 1941 was not prepared for the terrible conditions of unpaved Russian roads, and their transport became stuck in the *rasputitsa*.

Left to right: Admiral Raymond Spruance, who was appointed to lead the carrier attack on the Japanese at the Battle of Midway, and Admiral Isoroku Yamamoto, commander in chief of the *Kido Butai*.

The flight deck of USS *Yorktown* shortly after the first Japanese bombing attack of June 4 at Midway. The dense smoke is coming from a hit on the ship's funnel, but the radar and guns can still be seen on alert.

Above: British soldiers surrendering to a Japanese patrol following the capture of Singapore in February 1942.

Left: President Harry S. Truman *(left)* and General Douglas MacArthur at Wake Island, October 1950.

Left to right: General Henri Navarre, Colonel Christian de Castries, and General Rene Cogny reviewing the honor guard during inspection of the camp at Dien Bien Phu, February 1954. Two of the men can be seen limping on canes due to war wounds.

Viet Minh soldiers assaulting French positions on the Dien Bien Phu battlefield, April 1954. It is worth noting that this photograph—among others taken by the Vietnamese—is likely to be a reconstruction, taken a few hours after the actual event.

Rejoicing Viet Minh troops on a victory parade through Hanoi, October 1954.

remained to the Wehrmacht. Despite heavy losses over the past three and a half months of nonstop combat, this was still a more formidable force than that which had smashed the French defenses the previous year. Facing them, organized into three fronts, were some 1,250,000 Soviet soldiers, 3,232 tanks, 7,600 guns, and 936 aircraft (of which probably only 500 were still operational). Many of these troops would soon be battered and exhausted from their struggle to avoid annihilation in the impending Bryansk and Vyazma *Kesselschlacht*.

Starting on October 2, the first blows of Operation Typhoon struck home with devastating power. Moving up from the south to concentrate on Moscow, Guderian's Second Panzer Army conjoined with Kluge's Fourth to trap yet another large Soviet force in two vast pincer movements between the centers of Vyazma and Bryansk (some 150 miles west of Moscow and 220 miles southwest, respectively) in order to open the road to Stalin's capital. A total of five Soviet armies found themselves encircled in the Bryansk-Vyazma *Kesselschlacht*. Moscow's first line of defense had been shattered. By October 10, the Germans had captured another 673,000 Soviet prisoners. A quarter of a million Soviet soldiers had been killed or wounded in these two battles, in addition to the hundreds of thousands captured. Once again, Stalin was unable to believe what was happening. The air force commander who delivered this latest piece of bad news was arrested and handed over to the NKVD.

Stalin now had only 90,000 men and 150 tanks upon which he could draw for the defense of the capital. For the Panzers, the road to Moscow must at last have seemed open—weather permitting. But, once again, it would prove to be late—very late—in the campaigning season.

As if the summer had not been draining enough, the weather now turned against the invaders with a vengeance. At first the autumn rains brought relief by damping down the fearsome dust of the summer. Then, as they strengthened, they brought with them the curse of the autumn *rasputitsa*, the bottomless sticky mud that turned the Russian roads, the majority of them unpaved, into unnegotiable quagmires. A road-oriented force like the 1941 Wehrmacht was not prepared for these conditions. The Panzers with their narrow tracks sank out of sight. The six hundred thousand large western European horses that the Germans used for essential supply and artillery movement also floundered, while the smaller, tougher indigenous Cossack ponies of the Red Army were much better suited to the conditions. Photographs of that campaign show a Wehrmacht staff car being dragged through the *rasputitsa* quagmire by twenty infantrymen, and transport horses sunk up to their bellies in the glutinous mud.

If life was harsh for the invading soldiery, it was incomparably worse for the civilians caught up in this most savage and merciless of conflicts. Imbued with Hitler's doctrine calling for a war of extermination against *Untermenschen*, the invaders showed little compunction about turning inhabitants out of their homes to face death by starvation or cold, or tearing down their primitive dwellings to provide firewood. Meanwhile, the Russian civilians themselves faced draconian edicts requiring them to fulfill the policy of "scorched earth" warfare so that nothing should be left standing that might benefit the occupiers. It was a leaf taken out of the book of their ancestors of 1812, followed to the letter. As the Wehrmacht pushed ever farther east, it drove before it an ever-increasing tide of homeless refugees who had lost everything and had nowhere to go. That sublime chronicler of the misery of the war in Russia, Vasily Grossman, wrote at the time of the Vyazma-Bryansk debacle:

> I thought that I'd already seen a retreat, but I had never seen anything like this before, nor even imagined it. It was a Biblical Exodus! The vehicles were travelling in eight lines, engines screaming hysterically as dozens of lorries tried to get themselves out of the mud. People were driving huge flocks of sheep

> and cows over the fields, further off horse-drawn carts creaked along, thousands of carts, covered with coloured sackcloth, plywood, tin, carrying refugees from the Ukraine, still further ahead there were crowds of pedestrians with sacks, bundles, suitcases. It was not a stream, not a river, it was the slow movement of a flowing ocean, hundreds of yards wide from side to side. There were the black and white heads of children looking out from the covered carts, the Biblical beards of Jewish elders, the headscarves of peasant women, older men in Ukrainian caps, black-haired girls and women. And what calmness there was in the eyes, what wise grief, what a sense of Fate, of universal catastrophe!

Profoundly moved, Grossman added, "If we are to be victorious in this terrible, brutal war, it will be because deep in our nation we have people with these great hearts and souls, people who spare themselves nothing, like these old women, the mothers of these sons, who in their great simplicity 'lay down their lives for their friends,' as simply, as generously, as that poverty-stricken old woman of Tula, who gave us her food, her light, her firewood, her salt. They are like the righteous of the Bible. They illuminate all our people with a miraculous light. They are a handful, but they will be victorious." The fear that drove on this human tide was fanned by proliferating reports of the atrocities being committed by the Germans in Soviet territory, including the murder of thirty thousand Jews in the Ukraine. These had included Grossman's mother.

Grossman was to be proved right about the Russian spirit. But the Russian weather had further torments in store for the invaders. After dust and then mud would come snow.

CHAPTER 11

General Winter

ON OCTOBER 5, 1941, at his Leningrad general headquarters, Zhukov received orders from Stalin to return to Moscow at once and stabilize the front.

"Can you get on a plane and come to Moscow?"

"I request to fly at dawn."

"We await you in Moscow."

"I'll be there."

"All the best," Stalin said.

Stalin decreed that the wreckage of the reserve and western fronts be combined into a single western front under Zhukov. It was on this front that the main weight of the German assault on Moscow would fall. It was not difficult for Zhukov to deduce at once that the situation at the front was dire, more dangerous than any time since June 22. On October 3, Guderian, plunging deeply northward in brilliantly sunny weather, had taken the important center of Oryol, supposedly over a hundred miles behind the Russian front line. Thrusting on to Tula, he opened a southern flank directly menacing Moscow. The following day, the fourth, was the day that Stalin lost all contact with General Ivan Konev's shattered western front; he threatened to have Konev shot, and he probably would have done so but for a strong intervention from Zhukov. There was now a twelve-mile hole in the Moscow defenses. On the fifth, Soviet planes reported a long column of Panzers heading down the Ukhnovo highway, only sixty miles from Moscow. Passed was the famous battlefield of Borodino that had helped guard the route to Moscow in 1812. The following day the first snows fell. On the eighth, Zhukov, by now back in the capital, warned the Soviet High Command that the roads to Moscow now lay virtually undefended, and that the available reserves were "encircled."

It would be hard to think of a commander better suited to take over the defense of Moscow at this desperate time than Georgy Zhukov. Only forty-four years old, he seemed to have brought success to almost everything he had touched. After Nomonhan, having escaped from any blame for the Finnish debacle, he had appeared as the eleventh-hour rescuer of Leningrad in the late summer of 1941; his counterstroke at Yelnya in August and September, which had inflicted one of the first and heaviest reverses on the Germans, had delayed the German advance on Moscow for several crucial weeks. He was unafraid of heavy losses—on the contrary, he would go down in history as a butcher. Nor was he afraid of Stalin; those close to the scene were often stunned to see Zhukov fearlessly answering back, standing up to the Red Tsar. Surprisingly, "Stalin seemed to accept it as normal," noted one corps commander. Time and again, Zhukov would lash out at Stalin's call for costly and uninspired "frontal attacks," time-wasting ventures that would rob his offensives of their momentum. His most serious criticism was the overall lack of mobile formations.

Zhukov, writes his biographer Geoffrey Roberts, came across "as a general willing to execute the orders of his superiors without demur and who expected the same of those serving under him." His style of command was to seize the initiative by personal reconnaissance and bold offensive action, followed by "the skilful coordination of ground and air assets, and acceptance of heavy casualties if the situation demanded it." If the situation presented itself, he was keen to repeat the tactics that had succeeded so well against the Japanese at Khalkhin Gol, but on a much larger scale: namely, to pin down the enemy's main forces in the center "while powerful mobile forces attacked on the flanks, creating openings and encirclements that could be exploited by strong reserves." But this would have to wait until the balance of forces was more favorable, at Stalingrad late the following year.

Upon taking over command, at a moment when morale was at rock bottom, Zhukov had issued to all his forces on the Leningrad front his Order 0064 of September 17, 1941, decreeing that "all commanders, political officers [commissars], and rank and file who leave the line of defence without prior written instruction of the Front or Military Council are to be shot on sight." He could well assume in his new command that, if Moscow were lost, he would lose his head too. By late October, Zhukov's forces had succeeded in halting the German attack at least temporarily, and he could declare, "I will wear him [the enemy] out and then beat him."

On October 17, the Stockholm correspondent of the London *Daily Telegraph*, covering the Moscow front, reported a serious worsening of the situation. That the steely Soviet censors would permit transmission of news that the Germans "had at one point penetrated the line and reached the outer line of defences" only sixty miles west of the city, and that "both sides sustained heavy losses," indicated that things were going badly indeed. The key bastion of Mozhaisk (or Mozhaysk), the Borodino of 1812, had fallen, and a surge of panic swept through the city. Successive waves of refugees brought more tales of the Wehrmacht's brutality and its apparently irresistible impetus.

To keep up pressure on the embattled Muscovites, the Luftwaffe took to round-the-clock terror bombing of civilian areas. Although much damage and many casualties ensued, the raids could never reach the intensity of the Blitz on London the previous winter. Operating at extreme range from home bases, Göring's elite force was handicapped by a lack of munitions as well as inadequate repair facilities. And the Luftwaffe over Moscow found itself flying into flak even fiercer than that which had protected the much larger area of

London. Its limitations showed the weakness in the whole strategy of attempting to subjugate so vast a country with the means that had overpowered France in six short weeks.

That first panic at the beginning of October sparked looting and random killings. The dread scent of collapse even infiltrated into the inner recesses of the Kremlin. A "sinister snow" of black flakes from the ashes of documents being burned there drifted across the city. On October 19, Beria told Molotov, "We should abandon Moscow. Otherwise they will wring our necks like chickens." There must have been many thousands of Russians who would have been eager to do the wringing, if only they could reach Beria's own murderous neck. There were rumors, which still persist, that in early October Stalin put out feelers, via Bulgarian intermediaries, for an armistice. Orders were sent out from the Kremlin (which were invariably picked up by the shaky citizens of Moscow) to evacuate the government to Kuybyshev (also Kuibyshev), a city on the Volga five hundred miles to the east. At the same time, measures were set in motion to ship by the score whole factories eastward to the Urals. It was a staggering undertaking given the prevailing shortages of transport and manpower. Requiring forty carriages and thirty-five goods wagons alone, the entire Bolshoi Theatre and its artists were transported to Kuybyshev.

Travelling with the Bolshoi was the thirty-five-year-old composer Dmitry Shostakovich. He recalled scenes of panicky chaos as travelers trampled on one another in the slushy snow to get aboard trains at the Kazansky Station. His family struggled through the mob carrying in their hands a sewing machine and a child's potty. The journey took seven long days and nights, but when they reached Kuybyshev, conditions were as comfortless as they had been on the train, the family packed into one room with fourteen other people, all sleeping on the floor. Working on a new symphony, he had developed writer's block. "As soon as I got on that train something snapped inside me. . . . I can't compose just now, knowing how many people are losing their lives." But when he later heard of the German defeat in front of Moscow, "he finished the symphony in a burst of exuberant energy in less than two weeks." It was to become the renowned Seventh, or Leningrad, Symphony.

Another train, complete with a special refrigeration plant and shock absorbers, was employed to ship out Lenin's grisly mummy from its resting place in Red Square; this seems a curious priority for a city on the brink of defeat and with no rolling stock to spare. On the other hand, many of the innumerable arms and heavy manufacturing plants in Moscow were shipped east, lock, stock, and barrel, including the factory producing the key T-34, which—with extraordinary speed—was soon turning out this prize piece of fighting equipment from a new site safely in the depths of Siberia.

In his moving book *Moscow 1941*, Rodric Braithwaite graphically describes the remarkable exodus. One of those evacuated was a student at Moscow University who later became a world-famous nuclear scientist, the Nobel Prize winner and dissident Andrei Sakharov. Sakharov recalled students, in temperatures below -20°F (-29°C), stealing lumps of coal from the locomotive to fuel the stoves in their wagons, which housed forty in each car in appalling conditions. It took them six weeks to reach Ashkhabad in Turkmenistan. Often trains would arrive after journeys of several weeks to find that there were no buildings ready.

Nevertheless, the astonishing wholesale evacuation of industrial plants succeeded. By its end, writes Braithwaite, "five hundred factories and two million people had been moved from the city. It took over seventy thousand trains." The operation was completed by November 25—a miraculous achievement, and just in time for Zhukov's winter offensive.

The importance to the overall Soviet war economy can be understood from the statistics: by October, after only three months of war, the country had lost 62.5 percent of its coal production (largely in the captured areas of the Ukraine), more than two-thirds of its pig iron, steel, and rolled metals (71, 68, and 67 percent, respectively), and 60 percent of its aluminum. Before the war, half of the country's automobiles, half of its machine tools and instruments, and more than 40 percent of its electrical equipment had been produced in and around the capital. By November, as Stalin completed the evacuation of industries from Moscow, the loss of over three hundred factories to the enemy had deprived the Red Army of what had once been a monthly production of 8.5 million shell cases, nearly 3 million mines, and 2 million aerial bombs. The loss of chemical plants cut the output of explosives, causing a grave crisis because the Red Army had by then almost exhausted its prewar stocks of ammunition. The relocation of Moscow's factories was, as the war historian John Erickson notes, something "little short of an Industrial Revolution," carried out within a matter of weeks, an undertaking that itself created immense demands for fuel, raw materials, and construction labor.

As panic took over in October, there were regular reports of parachutists landing in the middle of Moscow. Its guards having fled, even the British embassy was looted. Backed by all the savage rigor of the NKVD, Zhukov introduced the most draconian measures. Martial law was declared. "Suspicious" citizens were arrested and shot without pretext. Years later, Beria would boast that, during the Battle of Moscow, 638,112 men had been detained and 505 "deserters" executed. On October 30, Zhukov issued an edict warning that "cowards and panic-mongers" fleeing the battlefield and "provocateurs, spies and anyone fomenting unrest" would be shot on the spot. "Not a step back! Forward for the Motherland!" was the cry. Special "blocking units" were introduced with orders to shoot any troops seen falling back without orders at once.

Military and police patrols appeared on the streets again. The panicky flight ceased. Ten thousand deserters and 25,000 people trying to evade military service were arrested. There were reports that, over the period from 1941 to 1942, a total of 157,000 Soviet servicemen were executed, the equivalent of more than fifteen divisions. According to Braithwaite, it was said that "some executions were carried out by a murderous woman in a shed in the courtyard of the Moscow City Court." On October 16, more than two hundred people were shot in Moscow by the NKVD, the biggest killing in a single day since the Great Purge of the 1930s. The lucky ones might escape immediate death by being sent to penal battalions charged with the task of clearing minefields in advance of an attack. Their life expectancy was not high. Another round in the purge of senior commanders took place. One of the Red Army's ablest tacticians, General Dmitry Pavlov, who had commanded the key western front in the first terrible weeks, the unfortunate general who on the eve of Barbarossa had been enjoying an evening out in Minsk, had already been executed in July, together with several other senior officers. Then, in October, among those liquidated was General Yakov Smushkevich, the commander of the Soviet air force, an intrepid Jew who, having fought through the Spanish Civil War, had been in charge of Zhukov's air power at the Battle of Khalkhin Gol, and made a Hero of the Soviet Union—now executed on the orders of Lavrenty Beria on October 28, 1941, the same day as Zhukov's predecessor General Shtern. There were suggestions of anti-Semitism in his death, an undercurrent never far from the surface in Stalin's Russia.

Yet even in a totalitarian state, victory could not be achieved through the rule of terror alone. The end of October must have looked like the eleventh hour to the frightened men inside the Kremlin, as well as to the general populace outside the wall. Morale had never been more fragile. Zhukov's reserves were almost exhausted. And still the Germans kept coming closer to Moscow. But, almost alone, Stalin refused to be panicked. Suddenly he managed to pull out of the hat a supreme trick. After all the disasters of the past four months, finally he made a decision of extraordinary courage, one that would go far toward earning him a reputation as a great war leader. He announced to his inner team that he intended to mount a full-scale military parade in Red Square on November 7 to celebrate Red Army Day, as had been done every year in peacetime. Members of the Stavka reacted with horror. The Germans were at the very gates of Moscow, a couple of hours' drive away, and what if the Luftwaffe, with its superiority in the air, were to bomb the parade? Stalin remained adamant. He consulted Zhukov, who assured him that the city would hold and that the enemy was "in no condition to organise a major attack in the next few days," and encouraged him to go ahead with the parade. Stalin did.

As in peacetime, Stalin took the salute atop the now empty Lenin Mausoleum, together with members of the Politburo who had not left for Kuybyshev. The Luftwaffe was conspicuous by its absence. The only thing that went awry was one tank that ran out of control and turned in the wrong direction. That drew a laugh even from Stalin. To hold the parade, Stalin had had to scrape the barrel to gather up cadets, veterans, and troops of an internal security division. They totaled nearly thirty thousand—or the equivalent of two large divisions. It was a demonstration of tremendous audacity. In a windy speech full of bombast, Stalin predicted that "within little more than a year, Hitlerite Germany will collapse under the weight of its own crimes." Needless to say, it was received with lengthy ovations.

What was particularly significant in his speech was his invocation of the example of "our great ancestors—Alexander Nevsky, Dimitry Donskoy . . . Alexander Suvorov, Mikhail Kutuzov"—all of them victorious generals from the tsarist era. It was an artful device for turning the war, switching from the existing propaganda image of the struggle of the Marxist-Leninist proletariat to one of historic nationalism. Such nationalism would inspire every patriotic Russian. Soon he would even be allowing, indeed encouraging, religion, that "opiate of the masses," to flourish once again, because that was what he sensed the ordinary Russian now wanted. Here was born the notion of the Great Patriotic War. Rodric Braithwaite sees the November parade as "a remarkable, perhaps foolhardy gamble. But unlike some of Stalin's other gambles in 1941 and 1942, it paid off in spades." Morale soared, as Muscovites seemed to regain faith in themselves.

Nevertheless, by the beginning of November the defense belt around Moscow was still dangerously incomplete. It had been hard to recover from the loss of 90 percent of the Red Army's tanks in the first week of the war, let alone the loss of three million in prisoners of war, and possibly another three million killed. One-third of Russia's rail network was behind enemy lines. War production was starkly diminished. Stalin was reduced to handing out packets of fifteen tanks at a time to threatened sectors.

With frenzied speed, Zhukov set about building concentric lines of defense around the capital. Some six hundred thousand citizens of Moscow—students, clerks, office workers, women, artists, and pensioners—were conscripted into building new defensive lines well into the city suburbs. At first they were dispatched without spades or pickaxes, with neither food nor accommodation. Typifying the wasteful inefficiencies of the Soviet system, little was achieved. Alongside these construction crews were raised hastily formed *opolchenie* units, a kind of home guard. All they had by way of arms were antique foreign rifles, and only one for every two volunteers. It was hardly surprising that the *opolchenie* produced no more than ten thousand recruits.

Under Zhukov's forceful drive, backed with menaces and threats, the Moscow construction teams gradually achieved extraordinary results with their limited materials. Great antitank ditches, some of them thirty feet deep and supported by tetrahedrons of steel, blocked all the routes entering the city—a tactic that recognized that the Panzers were essentially a road-borne force. Skillfully placed earth revetments were dug, which a T-34 could run into and fire from, leaving only its heavily armored turret visible. Nevertheless, with a maximum of just ninety thousand troops of mixed quality now left available to defend a

150-mile-wide front with open flanks at each end, Zhukov's force was stretched to the limit. As a political commissar remarked to a general, "We are generals without troops." Sometimes untrained units had to be thrown in piecemeal, at appalling cost; one newly arrived cavalry division from Tashkent, Uzbekistan, was massacred almost in its entirety, at zero cost to the enemy. Only a general with the standing and the reputation for ruthlessness of Zhukov could get away with such casualties. By the beginning of December, the German army commander Field Marshal Walther von Brauchitsch reported that the Red Army had "no large reserve formations"; it was a spent force.

If the Wehrmacht, in continuing to push relentlessly toward Moscow, still seemed like an unstoppable, superhuman force, how did the scene seem to the Germans in the front line? In conquering the whole of western Russia, German casualties had amounted to 46,000 by the end of July. Yet another 118,000 had died in the battles for Kiev and Leningrad and in the advance on Moscow. By the end of November, casualties would amount to 23 percent of the effective overall strength of the Wehrmacht, or nearly three-quarters of a million.

By the end of that same, unusually cold November, when the temperature had reached 40°F below zero (–40°C), Bock was reporting that his men had still not received winter coats; instead, they often had to pack newspapers into their jackets. Soviet soldiers, in contrast, had warm, quilted uniforms, felt-lined *valenka* boots, and fur hats. Hitler, in all his hubristic arrogance, had refused to consider that Barbarossa could possibly run on into the winter, and had effectively blocked cold-weather measures. Moreover, the Soviet partisans, by the end of 1941 increasingly well organized and fanatically committed as a consequence of German atrocities, were striking telling blows at the Wehrmacht's long lines of communication and supply deep behind the front. Essential supplies of food, clothing, fuel, and ammunition were simply not getting through. Soviet intelligence had noted, as early as the last days of September, that a German attack at Pulkovo, on the approaches to Leningrad, had been accompanied by no more than twenty tanks, which was taken as a sure sign of a slackening of effort.

This gave Zhukov a vital breather, at a moment of extreme crisis, in which to consolidate the defense structures and to organize the formations of newly arrived reserves. By now he had more or less completed three concentric defense lines circling the city.

On November 6, the day before Stalin's triumphal Red Army Day parade, it began to snow again, and then swiftly to freeze, sheeting roads and fields with a hard surface on which a highly mechanized army could move. The mud solidified too. The attackers could resume their deadly advance. Bock still had almost as many men as Zhukov's defenders; he also had many more guns, and more than twice as many tanks. In mid-November, the Germans' left hook managed to reach a point a mere fifteen miles north of the city. Their intention was to push the Third and Fourth Panzer Armies across the Moscow Canal, to envelop Moscow from the northeast. Guderian's Second Panzer Army would attack Tula and then close in on Moscow from the south, the second of two deadly German tentacles that were intended to reach around the rear of the city. As the Soviets fought the threat from the flanks, the Fourth Panzer Army would attack the center, straight down the highway to Moscow. Facing them, Zhukov had the Fifth, Sixteenth, Thirtieth, Forty-Third, Forty-Ninth, and Fiftieth Soviet armies, all of them fatigued and under strength.

On November 15, after a day's ferocious battle, Zhukov's Seventeenth Cavalry Division had eight hundred men left; three of his divisions were no longer in touch with army headquarters. The whole of his right wing was fighting desperately to hold off the growing threat of encirclement. In two weeks of relentless fighting, although lacking sufficient fuel and ammunition, the Germans slowly crept toward Moscow. By November 19, Lieutenant General Konstantin Rokossovsky found that his Eighteenth Army, which was blocking the vital Volokolamsk highway, was in danger of being outflanked on both sides. Zhukov ordered him to fight where he stood:

> Kryukovo [a village twenty-five miles from Moscow] is the final point of withdrawal: there can be no further falling back. There is nowhere else to fall back. All and any measures must be taken quickly to win a breathing space, to stop the retirement. Each further step backward by you is a breach in Moscow's defences. All commanders, from juniors to seniors, [are ordered] to be in their places, on the battlefield.

Rokossovsky, convinced that he should withdraw, appealed against this order. Zhukov threatened to have his onetime superior in rank shot for disobedience. The Germans reached Yasnaya Polyana, the home of the revered Leo Tolstoy, just southwest of Tula, on the road to Moscow. There were many heroic episodes; for instance, that of Major General Panfilov's 316th antitank men, ground to pieces while fighting German tanks on the Volokolamsk highway, where today a forty-foot statue still marks their stand. "Even Stalin's nerves were shaken" by setbacks in the defenses of Moscow at this point, records Braithwaite. To the south, the weight of Guderian's northward thrust struck two divisions on the Fiftieth Army's left flank, the 413th and 299th, the latter no more than eight hundred men strong, while the reserve, the 108th Tank Division, had two thousand men and thirty old T-26 tanks. However, the Second Panzer Army was blocked.

On the German side, General Halder was noting in his war diary that some German regiments were now led by an *Oberleutnant* (captain). On November 19, a whole German division was reported to have panicked. It was the first time that such a thing had happened. The Germans now had no more battalions in reserve; nor, apparently, did Zhukov.

Stalin now played his second ace of the campaign. German intelligence officers were suddenly stunned to discover their forward units being counterattacked with vigor by fresh and hitherto unidentified troops in large formations. They constituted the vanguard of some thirty divisions of Siberian troops just arrived from the Soviet Far East. They were tough and fresh, and seasoned with the battle experience of having fought the Japanese to a standstill at Khalkhin Gol just two years previously—under Zhukov. No troops in the world were better acclimatized to the unprecedently harsh cold that was about to grip the Moscow front. They arrived with their own spanking new T-34s, painted white, supported by assault troops clad in white sheeting—a terrifying phantom army, thrown into battle by Zhukov almost immediately after they decanted from the train after their long trip across the Trans-Siberian Railway. These Siberian reinforcements arrived just in time to reverse the tide of battle. Augmenting Zhukov's hard-pressed Forty-Ninth and Fiftieth armies, they struck a savage and totally unexpected blow on November 22 against Guderian's Second Panzer Army, which was attempting to encircle Moscow from the south. This attack inflicted a shock defeat on the Germans, a reverse on a scale that had not been seen since Barbarossa began. Bock wrote that he now saw himself facing "a Verdun": "a brutish, chest-to-chest struggle of attrition."

However, the Fourth Panzer Army still managed to push the Soviet Sixteenth Army back, splitting it from the Thirtieth, and succeeded in crossing the Moscow-Volga Canal. From the northern tentacle, the Third Panzer Army finally captured the important town of Klin after heavy fighting on November 24. Zhukov recalled that it was during this time that Stalin asked him, "Are you sure that we can hold Moscow? I ask you this with pain in my soul. Speak honestly, like a communist." Zhukov replied, "We can certainly hold Moscow, but at least two more armies will be needed, and if possible two hundred tanks."

By November 28, the German Seventh Panzer Division (the veteran division that General Erwin Rommel had led in the French campaign) had seized a bridgehead across the Moscow-Volga Canal, bringing them less than twenty-two miles from the Kremlin, before the Russian First Shock Army forced it back across the canal. Moscow was in sight, but the German forces were wearing out.

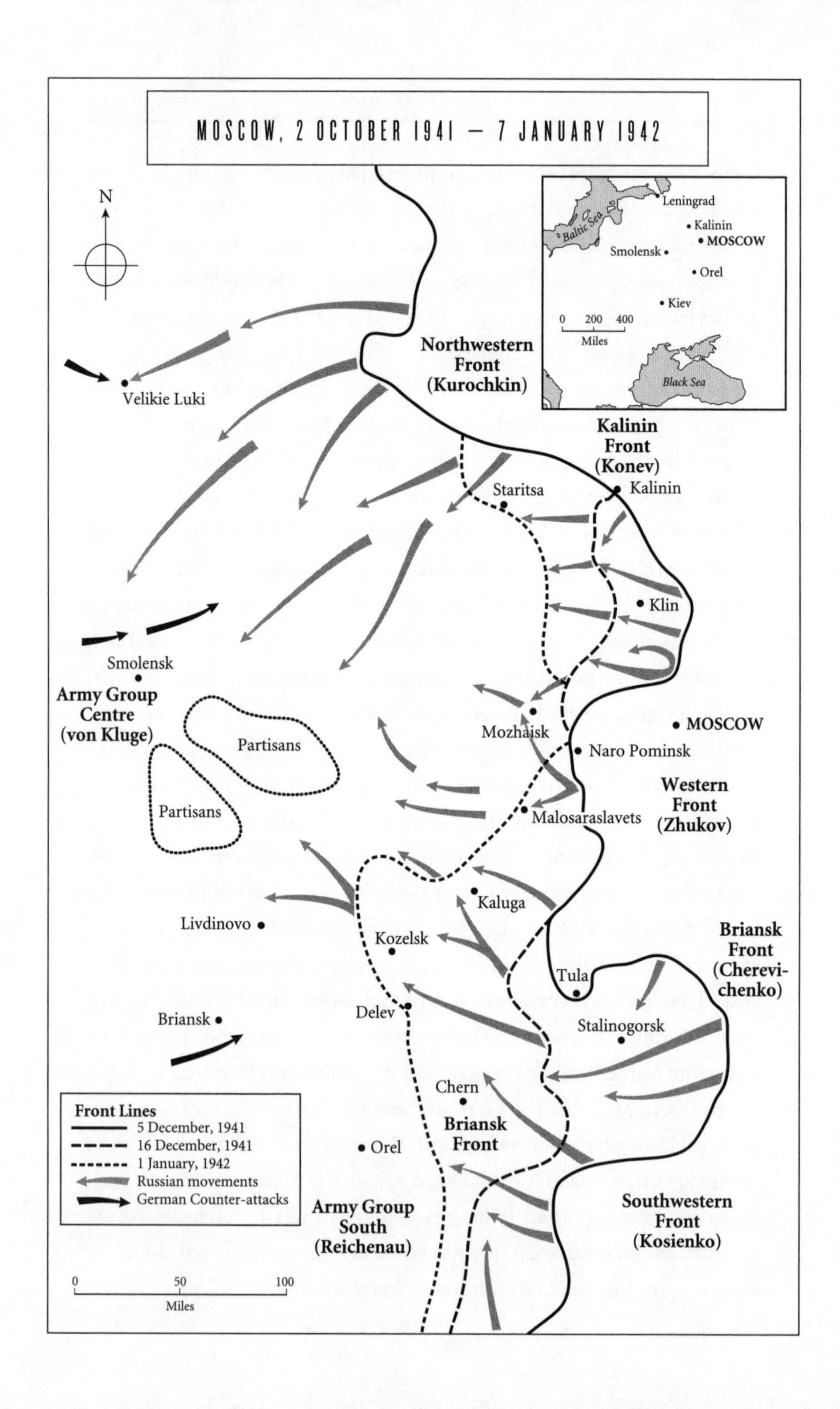
MOSCOW, 2 OCTOBER 1941 – 7 JANUARY 1942
N
Leningrad
Baltic Sea
Kalinin
MOSCOW
Smolensk
Orel
Kiev
0 200 400
Miles
Black Sea
Northwestern Front (Kurochkin)
Velikie Luki
Kalinin Front (Konev)
Staritsa
Kalinin
Klin
Smolensk
Army Group Centre (von Kluge)
Partisans
Partisans
Mozhaisk
MOSCOW
Naro Pominsk
Western Front (Zhukov)
Malosaraslavets
Kaluga
Livdinovo
Kozelsk
Briansk Front (Cherevichenko)
Tula
Briansk
Delev
Stalinogorsk
Chern
Briansk Front
Orel
Front Lines
5 December, 1941
16 December, 1941
1 January, 1942
Russian movements
German Counter-attacks
Army Group South (Reichenau)
Southwestern Front (Kosienko)
0 50 100
Miles

On the southern approaches to the city, Zhukov managed to hold the key city of Tula on the tip of Guderian's prong; the Wehrmacht never got closer to the capital from this direction. Frustrated by Soviet resistance on both the northern and southern approaches, Bock attempted another offensive from the west on December 1, but it was soon repelled. On December 2, one German reconnaissance battalion actually managed to reach the Khimki railway station, only eleven miles from the center of Moscow. There its men could clearly see the spires of the Kremlin. It would be the high tide of the German attack on Moscow. (Tourists arriving by air in Moscow today will be shown the monument marking this spot; it stands close to Sheremetyevo International Airport, now well inside the city's suburbs.) A turning point in the battle—if not, with hindsight, a turning point in the whole war—had been reached. By now the German attack was beginning to peter out. The Russian winter was about to add its weight to the balance.

So how had Stalin managed to acquire this vital infusion of fresh troops that would tip the precarious balance between Zhukov and Bock and save Moscow? It was the consequence of a piece of intelligence from the agent whom earlier in the year Stalin had condemned as a "drunk and a pimp": Richard Sorge, in Tokyo. Despite the disheartening rebuff he had received for his highly accurate messages about the date of Barbarossa, Sorge continued to supply his masters in Moscow with top-grade intelligence throughout the summer of 1941. What Stalin wanted to know from him, above all else, was whether there was any likelihood that Japan might seize on the Soviet reverses in Barbarossa to launch an attack in the Far East, in revenge for the defeat at Nomonhan. As early as August, Sorge had been reporting that this was unlikely because Japan's proponents of "go south" had won the day and were busy preparing the strike on Pearl Harbor. (This was not, however, a piece of information that Stalin would pass on to the United States, which he continued to distrust at least as much as he distrusted Nazi Germany.) Later that month, Sorge sent in a very detailed report stating emphatically that Japan would not attack the Soviet Union until Moscow had been captured, "the Kwantung Army's troop strength [had become] three times that of the Red Army," and "signs of internal collapse in the Siberian Army" had become visible.

From his travels in Manchuria, Sorge's agent Hotsumi Ozaki had passed to Sorge confirmation that "the Kwantung Army has decided not to have a war with Russia." This was all so specific and authoritative that even Stalin could not disregard it, though Sorge would claim that it was not until the end of September that Moscow finally did believe this report.

Thus, at a critical moment when the final, all-out German drive on Moscow was under way, Stalin took the calculated risk of dispatching those highly qualified troops from Siberia—nearly a million in total—to reinforce Zhukov, their onetime commander. They would arrive just in time to tilt the balance.

As well as being perhaps the most valuable piece of intelligence Sorge would ever send, it was also to be his last. Stalin's cold-shouldering in the spring had imposed a strain on Sorge's nerves at an increasingly difficult time, just as the Japanese secret police were moving in on him. He took to drinking more heavily than ever, racing around Tokyo on his motorbike and throwing caution to the winds. The beginnings of the roundup of his ring had begun. Miyagi threw himself out of a police station's window, having come clean about Sorge. Sorge sent his last message to Moscow on October 4, his forty-sixth birthday, and he was arrested fourteen days later, together with Max Clausen. Apparently feeling that his mission had been completed, Sorge gave a full confession (according to Gordon W. Prange's account, he was not tortured—perhaps surprisingly, given Japanese predilections).

The trials of these agents were held through the summer and autumn of 1942, while Stalingrad was under siege, and continued until August 1943. The Japanese made three overtures to the Soviets, offering to trade Sorge for one of their own spies. However, the Soviets declined all the offers, insisting that Sorge was unknown to them. Some historians have argued that Sorge's Barbarossa coup led to his undoing, because Stalin could not afford to let it become known that he had rejected warnings of the German attack in June 1941. Surprisingly, too, Sorge was kept alive in prison, and apparently treated reasonably well for another three years, only to be hanged on November 7, 1944.

It was not until the Khrushchev era, in 1964, that the Soviet Union officially acknowledged this extraordinary, German-born hero of the Great Patriotic War. The Soviet press claimed that, when he was about to be executed, Sorge had shouted, "Long live the Communist Party, the Soviet Union and the Red Army." It is hard to see how they would have acquired this piece of information. On November 5, 1964, Sorge was awarded the supreme decoration of Hero of the Soviet Union. A fulsome citation read:

> In this grave hour for our country . . . Richard Sorge and his fearless comrades once again rendered invaluable service to the Soviet people. They reported that the Japanese militarists, confident that the Hitlerites would cope with the Red Army, were concentrating forces for unleashing war in the Pacific. This information made possible the transfer of Soviet divisions from the Far East, although the presence of the Kwantung Army in Manchuria necessitated the Soviet Union's keeping a large number of troops on the eastern borders . . .

A street in Moscow remains named after him, and a postage stamp, valued at four kopecks, was printed in his memory.

At the end of November 1941, the freeze that, earlier in the month, had hardened the ground, thus briefly enabling Hitler's Panzers to launch their final bid for Moscow, now turned fiercely against the Germans. The mercury dropped to a record –58°F (–50°C), the coldest in Russian history. Nothing like this had ever been experienced by Napoleon. The Wehrmacht was not equipped for winter warfare. It lacked winter-grade lubricants; tank engines froze solid, motor transports were not better off, and even machine guns jammed in the cold. Horses died; so did the men. More than 130,000 cases of frostbite were reported, and Guderian signaled he had twice as many cases as battle casualties. Frozen grease had to be scraped off every shell before loading, and vehicles had to be heated for hours before use. Weather paralyzed the Luftwaffe. For the first time since the catastrophe of June 22, Russian aircraft were able to achieve something like local superiority. By November 27, General Eduard Wagner, the quartermaster general of the German army, reported, "We are at the end of our resources in both personnel and material. We are about to be confronted with the dangers of deep winter." The Soviet troops, particularly the Siberians, were much better prepared for subzero warfare. And the T-34s, specially winterized, remained dangerously active.

It hardly strains the imagination to picture the Wehrmacht camp in the last days of November as it strove to cover the last miles to Moscow. The young Germans in Bock's spearhead, now 1,500 miles or more from the *Gemütlichkeit* of home, found themselves in a frozen, dark, inhospitable land, whose every inhabitant was possessed by one wish only: to kill the invaders. Cold and hungry, their winter clothing not yet arrived or destroyed by partisans hundreds of miles behind the lines, and the local villages burned to the ground on Soviet "scorched earth" orders scrupulously carried out, the Germans were left with neither shelter, food, nor fuel. They watched daily the agony of the precious horses, essential even to a force as mechanized as the Wehrmacht; they died by the thousands, often left in their harnesses, bogged down inextricably in the *rasputitsa* mud or, later, frozen to death. At the same time they saw their Panzers simply worn out and not replaced, whereas a knocked-out T-34 always seemed to be replaced by three more.

So what kept the Wehrmacht soldier going? There were the two words fundamental to the German soldier over the ages: *Kameradschaft* (camaraderie: you never let your comrades down) and *Eidespflicht* (the duty that bound you to the military oath). Both were reinforced by fear—by fear of death at the front, and now by fear of the power of the secret police set up by the dictator who had dispatched the German soldier so deep into Russia. Yet there was manifestly another factor: a sense of exultation as forward reconnaissance units reported being within sight of the Kremlin's spires and towers. Here at last Bock's irresistible might was within reach of the holy grail. Moscow's fall, so Goebbels repeatedly told the *Frontsoldaten*, would inevitably be followed by the fall of Stalin, by the crumpling of the whole Soviet state—and, in consequence, by an end to the war.

The phenomenally icy weather continued into December, which saw some of the lowest mean temperatures ever recorded, day after day. The ill-equipped Wehrmacht, prepared by the self-assured Führer for a short summer campaign only, found it impossible to carry on. The attempt to capture the Soviet capital ground to a halt. "The offensive on Moscow failed . . ." General Guderian admitted in his diary. "We underestimated the enemy's strength, as well as his size and climate. Fortunately, I stopped my troops on 5 December, otherwise catastrophe would be unavoidable."

This was not, however, how Zhukov intended to let matters rest. At the end of November he submitted a bold plan to Stalin for a major counteroffensive. Stalin agreed. The forces for the attack were cobbled together from all the available reserves in the Red Army, and with a backbone provided by yet more fresh troops arriving in Moscow from Siberia. The units available to Zhukov numbered over five hundred thousand men, together with reportedly one thousand tanks and one thousand aircraft. On December 5, the very day that Guderian issued his halt order, they launched a massive counterattack. Their main objectives, sensibly, were the two threatening German prongs to the north and south of the capital. To great advantage, Zhukov employed the very Russian skill in *maskirovka* that had served him so well in Nomonhan against the Japanese. With negligible sources of intelligence available to them from a thoroughly alienated populace, Bock's spent forces were taken completely by surprise. Western armies, whose officers had been taught at Prussian Staff College, did not attack in the winter. They rested up. Certainly they would not think of moving in subzero weather. Anyway, the Red Army's losses had surely been so great as to make any offensive out of the question.

Zhukov's counterattacking infantry appeared clad in white camouflage capes, mounted on propeller-driven sleds and carrying tommy guns. Even the T-34s were painted white; all that could be seen of them were great spumes of snow as they rushed forward. To the Germans, the attacking Russians must have seemed like a great phantom army arising out of nowhere. Surprise was complete. Many of the T-34 squadrons had been formed as much as a hundred miles behind the front in some sectors, proving that mobile warfare was still possible in the Russian winter. All along the Moscow front German troops retreated. Soon after, the High Command in Berlin ordered a halt to all offensive operations. Bock wrote in his diary, "The Russians, who have destroyed almost all facilities on the main roads, have been able to obstruct our transport arrangements to such an extent that the Army Group no longer has what it needs to survive and to fight. . . . Today we no longer have the possibility of significant manoeuvre."

By December 13, German forces had retreated more than fifty miles from the capital. On the eighteenth, Bock was relieved of his command. Army Group Center, under his command, had come the closest the German army ever would to Moscow. Never again would the Soviet capital be threatened. To his diary Bock confided:

> All along, I demanded of Army High Command the authority to strike down the enemy when he was wobbling. We could

> have finished the enemy last summer. We could have destroyed him completely. Last August, the road to Moscow was open; we could have entered the Bolshevik capital in triumph and in summery weather. The high military leadership of the Fatherland made a terrible mistake when it forced my army group to adopt a position of defence last August. Now all of us are paying for that mistake.*

Over the next four weeks the Germans would be pushed back over two hundred miles. By then Barbarossa had cost them over 210,000 killed and missing and 620,000 wounded, a third of whom had become casualties after the beginning of Operation Typhoon. On top of this came an unknown number of Axis casualties such as Hungarians, Romanians, and Finns. On December 5, Hitler ordered his troops to assume a defensive stance. Guderian believed that the current front line could not be held, but Hitler was adamant: there could be no withdrawals. Guderian objected and was soon dismissed, as were a number of other generals including Bock for "medical reasons" (though he was brought back a few months later). Some forty other high-ranking Wehrmacht officers were also relieved of their command.

The Soviet offensive continued for the better part of a month, with Zhukov attempting on a few occasions (and with varying success) to repeat the double encirclement of Nomonhan against the Germans. Yet by the time Zhukov's offensive ceased on January 7, 1942, the Moscow front had been pushed back up to one hundred miles along six hundred miles of front. Moscow was now safe, and the Germans had been dealt their first major defeat of the Second World War. The myth of the invincible blitzkrieg had been shattered. It also provided a most important boost for Soviet morale. Hitler now had to prepare

*Bock was killed in the last days of the war, by a British fighter strike, the only one of Hitler's field marshals to die in action.

for a prolonged struggle. Operation Barbarossa had failed, stalled on the approaches to Moscow.

Hitler had conceived a low opinion of his regular Wehrmacht generals at various points in the victorious French campaign of 1940, when they had shown themselves to be indecisive or overcautious. It was only his superior will and intuition that had carried them through to victory—or so he persuaded himself. Now all his misgivings had been given new shape by the failure to seize the enemy capital, which had been so clearly within his grasp by the beginning of December. From his distant aerie, success was clearly just a matter of one more push. Then he would be recognized as the greatest warlord the world had yet produced. On December 19, Hitler took over the direct command of the German army from Walther von Brauchitsch. One of his first utterances was, "The will to hold out must be brought home to every unit!" That meant that no tactical withdrawal, however militarily desirable, would henceforth be countenanced. It would bring doom to German armies from Stalingrad to Berlin.

Buried in his East Prussian forest hideout, the Wolf's Lair, accessible only by his terrified toadies, Hitler sat completely out of touch with his frontline generals. His vision of the world was comparable to that of a Panzer driver, closed down in his tank, able only to peer through a narrow slit. He would vent his contempt and rage on the generals after every check and setback encountered by the Wehrmacht. News filtered through to him very slowly, and was usually tinged with false optimism. General Halder would write up in his diary all his fears, but he would seldom dare broach them to the Führer. Consequently, Hitler fed on the first (Japanese) tidings of Pearl Harbor, and believed that the American fleet was completely destroyed.

Hitler's conviction that he was a superhuman, however, now led him to commit one further act of supreme hubris. On December 11, 1941, four days after Japan struck at Pearl Harbor, he saw fit to declare war on the United States. Why? He did not need to; he was not bound by any commitment to the mikado. Inexplicable as the act still seems, it could only have come from some self-sown belief that, with the fall of Moscow only one push away, and with it the final conquest of Russia, he was about to become lord of the universe. Why should he share his glory with a contemptible Asian country that, in a matter of days, and with small losses, seemed poised to conquer a far larger empire than had Hitler and his mighty Wehrmacht? Why not seize this moment to display his contempt for the wobbling Anglo-Saxon powers?

After the great triumph of the winter offensive of December 1941, Stalin, too—overinflated by success—now went on to commit another of *his* calamitous mistakes. Encouraged by Zhukov's winter triumphs, he concluded that the Red Army was stronger than it really was, that one great push across the board would finish off the invaders. Accordingly, he ordered a general offensive for the spring of 1942. Zhukov, knowing just how exhausted his troops were from the winter offensive, was horrified. In the event, the spring offensive proved an almost unqualified disaster. The Wehrmacht was still too strong. Masters of defense, the Germans had reinforced the lines they had retreated to in December with a system of "hedgehog" fortified strongpoints. The Russians exhausted themselves battering against these, resulting in terrible casualties. Thus, when Hitler opened the new campaigning season with a thrust toward Stalingrad and the oil fields of the Caucasus, the Russians had few fresh troops in reserve to meet the new threat. The new German summer offensive of 1942 split the Russian front into fragments on either side of Kursk. Three armies (Weich's Second Army, Hoth's Fourth Panzer, and Paulus's Sixth Army), along with eleven Panzer divisions, fanned out toward Voronezh and the Don River. Paulus's Panzer divisions reached the Don on both sides of Voronezh on July 5, approaching the fringes of the oil-rich Caucasus—and ultimately Stalingrad.

That August, Zhukov was sent to assume the overall command of the sagging Russian front at Stalingrad. There he found a great gray tongue of German troops licking toward the River Volga at Stalingrad. Zhukov, the victor of Moscow, went with orders to command and coordinate the battered defenders, whose backs were to the river. As the enemy troops poured into the tongue of occupied steppe, he spotted an opportunity. Occupying a huge area, the overstretched Wehrmacht had long and highly vulnerable flanks. Both north and south of the city, these flanks were defended by Romanian or Hungarian Axis troops: troops of indifferent quality and resolution, and poorly armed. This was reminiscent of the situation in Nomonhan, where Zhukov had found flanks held by vulnerable and second-rate Manchukuoan troops. He thus applied the same tactics as at Nomonhan, though on a vast scale in which millions of troops, as opposed to tens of thousands, were involved. The ultimate result was the colossal defeat and total destruction of Paulus's army at Stalingrad. Over ninety thousand Axis troops were marched into captivity, from which only a few thousand ever returned. It was the disaster that marked an irreversible turning point in German fortunes in the Second World War. This was the battle with which Zhukov's name would be associated forever after. However, it should be stressed that Stalingrad would almost certainly not have happened had it not been for Zhukov's victory in front of Moscow. Rodric Braithwaite notes in his judicious account:

> By one measure—the number of people involved—the Battle of Moscow was the greatest battle in the Second World War, and therefore the greatest battle in history. More than seven million officers and men from both sides took part, compared with the four million who fought at Stalingrad in 1942. . . . This was a scale never matched in the fighting in Western Europe and Africa. The Battle of Moscow swirled over a territory the size of France, and lasted for six months. . . . The Soviet

> Union lost more people in this one battle—926,000 soldiers killed, to say nothing of the wounded—than . . . the combined casualties of the British and the Americans in the whole of the Second World War. This was the horrendous price they paid for inflicting on the Wehrmacht the first real defeat it had ever suffered.

These are statistics impossible to dispute.

There remains one crucial issue, still much argued about among historians: If Moscow had fallen in the winter of 1941, would the Soviet Union have collapsed with it? My personal view is that it probably would have; apart from the city's crucial importance as a communications hub, and—still, despite the evacuations—as an industrial complex, its psychological significance as the capital was unquantifiable, as indeed was Stalin's all-out commitment to its defense. Finally there was the proviso in Sorge's last message: if Moscow fell, Japan might be tempted to enter the war for some cheap gains in the Far East. Such a two-front threat would have demanded the certain return to Siberia of those formidable troops with which Stalin had turned the tide at Moscow, leaving the rest of Russia feeble prey for the Nazi hordes.

Could Hitler, then, have taken Moscow? Yes, if he had started Barbarossa earlier, and thereafter concentrated all his overwhelming force on that one objective. What a shattering reversal of arms for a hitherto omnipotent warlord, so steeped in hubris, and all its self-deluding conceits!

PART FOUR

Midway, 1942

CHAPTER 12

The Kido Butai

AT 7 A.M. on Sunday, December 7, 1941, Admiral Isoroku Yamamoto's carrier planes struck Pearl Harbor, perhaps the most written-up event in military history. Whether, as Japan claimed, a proper declaration of war got mislaid in the post is immaterial.* America was caught thoroughly by surprise, which of course made a day of infamy that much more infamous. As we have seen in part 1, Pearl Harbor should perhaps have been better prepared for such a sneak attack in the light of what had happened at Port Arthur in 1904, when the Japanese attacked without declaring war and while diplomatic negotiations were still under way. But history's lessons are too often ignored.

In the five months following Pearl Harbor, Japan swept the board in the Pacific, contrary to US estimates that it would not be capable of mounting more than one major naval operation at a time. In short order, Hong Kong, the Philippines, Singapore, the Dutch East Indies, and Burma were seized, and the seemingly irresistible Japanese were knocking on the doors of India and Australia. In Malaya and Singapore, bicycle-riding shock troops trained by that evil genius Colonel

*Yamamoto's original plan stipulated that the attack should not commence until thirty minutes after Japan had informed the United States that peace negotiations were at an end. In 1999, however, a professor of law and international relations in Tokyo, Takeo Iguchi, discovered documents that pointed to a revealing debate inside the Japanese government and which in his view showed "that the army and navy did not want to give any proper declaration of war, or indeed prior notice even of the termination of negotiations . . . they clearly prevailed."

Tsuji in French Indo-China, rounded up some 130,000 British and Australian troops with minimal losses. In his history of the Second World War, Churchill called the fall of Singapore to the Japanese the "worst disaster and largest capitulation in British history." Mopping up British naval units in the Indian Ocean, together with the surviving remnants of the Dutch fleet, the Japanese admiral Chuichi Nagumo had forced the abandonment of Ceylon (today's Sri Lanka) as a naval base. For the first time in its imperial history, the Royal Navy had to fall back on East African and South African ports, abandoning everything east of Durban, South Africa.

As the historian Samuel Eliot Morison notes, only one major battle in which the US Navy was involved "was decided by gunfire. All the others were won by ships' torpedoes, and by bombs and torpedoes delivered to the target by aircraft." The true queens of the chessboard would prove to be the newfangled aircraft carriers. As of December 7, 1941, Japan had six fleet carriers, whereas the United States had only two that were battle-worthy, to cover the whole Pacific. A storm had providentially prevented Admiral William F. Halsey Jr. and his *Enterprise* task force of aircraft carriers from reaching Pearl Harbor on the seventh.

Yet far more powerful than the loss of life and all those battlewagons was the psychological effect the attack had across the width and breadth of the United States. As Yamamoto is sometimes supposed to have said after the attack "All we have done is to awaken a sleeping giant and fill him with a terrible resolve." Without warning, America's own territory had been brutally violated; 2,402 men had been killed and 1,282 wounded. The backbone of the navy's battle fleet appeared to have been destroyed.

Heartless as it may sound, however, Pearl Harbor was in retrospect the best thing that could have happened for the Western alliance. After Hitler's attack on Poland in September 1939, and until the fall of France nine months later, Britain had fought a desultory and half-hearted war. The mighty French Army did absolutely nothing to help the poor Poles. But the United States was galvanized universally to total war after Pearl Harbor in a way that it had never been before, and never would be again. From being a nation at least half of whose people were dedicated to peace and isolation, America at once became passionate for war and victory. Even the commitment to global war, coming as it did four days later, with Hitler's declaration of war, was greeted almost with relief, and certainly with determined acceptance.

With it went an angry, racial hatred of the Japanese. The fighting in the Pacific would be fought with a savagery not seen on the western front in Europe. As the historian Alan Nevins puts it, "Probably in all our history no foe has been so detested as were the Japanese." It was primarily "the treachery of Pearl Harbor" that could never be forgiven, followed by the atrocities to the POWs on Bataan, and the murder of British nurses in Hong Kong and Singapore. Altogether it produced a "rage bordering on the genocidal," says John W. Dower in his excellent account of the struggle in the Pacific, *War Without Mercy*. It was exemplified by Admiral William F. Halsey's "Kill Japs! Kill Japs! Kill more Japs!" Such slogans would appear at home in movies or in the press, along with such sentiments as "Now you are going to get it—and it won't be finished until your detestable empire is wiped off the face of the earth." As the war correspondent Ernie Pyle remarked, "In Europe we felt that our enemies, horrible and deadly as they were, were still human . . . but out here [in the Pacific] I soon gathered that the Japanese are looked upon as subhuman and repulsive."

This informed the personal conduct of the war. In the South Pacific the psychology became "kill or be killed." No prisoners were taken, and no quarter on either side was given. Often Japanese naval survivors would be shot in the sea, sailors uttering the words, "We can't be sporting in a war." Harsh treatment was also given to the wretched native Japanese living in the United States, the "Nisei" (second generation), most of whom had been loyal citizens for many years. On the American West Coast, first- and second-generation citizens of Japanese ancestry were rounded up and committed to internment camps. As previously noted, even the highly respectable Smithsonian Institution in Washington, DC, came out with such unscientific remarks as "Japanese skulls are 2,000 years less developed than ours." As Dower notes, "the end results of racial thinking on both sides were virtually identical—being hierarchy, arrogance, viciousness, atrocity and death." This was what a combination of Pearl Harbor and the twisted legacy of Bushido had led otherwise civilized nations to. It was Japan's historic misfortune that American "hate" would possess the ultimate weapon of vengeance, the atomic bombs dropped on Hiroshima and Nagasaki. Well might one wonder how the chivalrous Admiral Togo would have reacted had he been able to witness the way his noble victory of 1905 would become debauched in such a vicious war.

Yet even amid the intense gloom that followed Pearl Harbor, for Americans—being Americans—there was always that spark of hope. That spark was turned into a flame when, on April 17, 1942, a small force of sixteen United States Air Force B-25 bombers, under the command of Lieutenant Colonel James H. "Jimmy" Doolittle, struck back at the very heart of Japan—one of the most courageous, self-sacrificing, and inspiring feats of arms undertaken by any Allied unit in the Second World War. It was just a pinprick, but it would hold important strategic implications for the subsequent course of the Pacific War.

Only two weeks after Pearl Harbor, Roosevelt summoned the Joint Chiefs of Staff to the White House to demand that a bombing raid on Japan be carried out as soon as possible, to restore the battered public morale. The carrier decided upon was the 27,900-ton *Hornet*. A lightweight by comparison with Japan's *Akagi* and *Kaga*, both of them over 41,000 tons, *Hornet* was a brand-new ship that was currently undergoing sea trials off the Virginia coast. She had a green crew, many aged no more than eighteen, recruits fresh from boot camp, some of whom had never seen the ocean. On February

2, 1942, they were stunned to see two experimental B-25s loaded on board. Then, after setting out to sea, they watched them take off—the first ground-based medium bombers ever to do so in the history of aviation. They were unable to return. Having completed their bombing mission, the B-25s were to overfly Japan and land on airfields in Russian Siberia or in China. Although Soviet Vladivostok would have been the nearest field, Moscow—fearful of getting embroiled in war with Japan—refused permission to land there. The additional mileage imposed in reaching China meant that the bombers would be straining for their last drops of fuel.

It was, however, the Chinese who suffered the most from the barbaric Japanese retribution that followed the Doolittle Raid. An estimated 250,000 Chinese from the areas that had helped the US airmen are reported to have been executed by way of reprisal. This horrifying figure has never, to my knowledge, been disputed; if accurate, it must make this act the most savage reprisal in all of the Second World War.

Of course, all sixteen of the valuable B-25s were lost. Ten thousand navy personnel were involved in the operation. Two of Halsey's indispensable carriers—the *Hornet* and *Enterprise*—were put at risk. So was it all worth it? In the United States, battered after months of unrelenting bad news from the Pacific, the effect on morale was tremendous. Here, for the first time, America was hitting back at the very heart of the Japanese war machine. The raids' effect was enhanced by Roosevelt's mysterious declaration—most alarming, by implication, to the Japanese—that the raiders had come "from Shangri-la." It would be twenty-six months until American bombers would strike again at Japan. By then, following the development of the B-29 Superfortress, each carrying ten times the bomb load of Doolittle's B-25s, the results would be devastating—culminating in the dropping of the atom bombs on Hiroshima and Nagasaki.

The Doolittle Raid had a powerful influence on Yamamoto's thinking. He warned that the raid could be a "taste of the real thing" that was to come. It meant that the defensive island perimeter that Japan had created around itself in those first heady weeks after Pearl Harbor was still not large enough. To make Japan safe, it would surely have to be pushed still farther out. Admiral Nagumo's raiding force, which had been sweeping the Indian Ocean as far west as Ceylon, was summoned home. Fighter units that had been earmarked for the Solomon Islands and the drive on Australia were withdrawn to protect the homeland.

More decisive in the war as a whole was the effect that the raid had on Japanese plans for attacking Midway Island (really an atoll embracing two islands)—the US base nearest to Japan. Two weeks before the Doolittle Raid, when Yamamoto presented his plan for the seizure of Midway, its feasibility had met with strong criticism from the army. Now, persuaded by the potential threat to the homeland, the doubters came around to support Yamamoto's Midway plan, and the operation was brought forward with urgency and excessive haste, to be launched at the beginning of June 1942.

In the spring of that year, the tide of war was running strongly against the Allies. In the Pacific, the seemingly invincible Japanese were threatening India and Australia. Never had an empire been acquired in so short a time, and certainly never one with such a huge perimeter. Japan's quest for raw materials, especially those essential to the unending war in China, now looked assured. The guarding outposts that protected that empire were being swiftly and powerfully entrenched, and reinforced, by Japanese arms. Gained in but a few weeks, how many months if not years would it take to roll it up again? Allied planners must have asked themselves this question countless times in head-scratching despair. How long before the great base of Singapore, captured by the Japanese in weeks, would once again fly the Union Jack? And where now were the bases from which such a return could be made? There were few enough of these left, now that the juggernaut of Japanese aggression had been at least temporarily halted in its relentless momentum. At one end there was distant Australia, plus a handful of islands in the Solomons and a tenuous grip on the southeastern tip of inhospitable New Guinea. At the other end, the tiny atoll of Midway, a solitary US outpost on the way to Japan, 2,600 miles away—and 1,300 miles distant from Pearl Harbor. For Japan, Midway was an essential stepping-stone to the capture of Hawaii, itself particularly vulnerable to invasion, and a threat to the West Coast of the United States. For America, for those very same reasons, Midway had to be fought for, and held, at any cost.

At the far west of the sprawling Hawaiian archipelago, at its hardest part to defend, the tiny atoll of Midway measures less than twice the area of New York's Central Park. In early May, Japanese and US carrier fleets fought out a bitter engagement in the Coral Sea, between the Solomons and New Guinea. It was the first naval battle ever to be fought in which the opposing fleets were never in sight of each other. American losses were heavier, in that they lost the carrier *Lexington* (more through American carelessness than enemy skill), and the *Yorktown* suffered heavy damage (in Tokyo, it was reported as sunk). But the Japanese lost forty-three aircraft and their hard-to-replace veteran crews. Most significantly, the drawn battle persuaded the Japanese to abandon their plans to seize New Guinea's Port Moresby, the jumping-off point for an invasion of Australia.

Nevertheless, the early summer of 1942 and the year's run of amazing successes presented the Japanese warlords with almost a surfeit of strategic options. In the prevailing state of Allied debility, most of these options were enticing ones—and all of them involved a naval strategy. The Japanese could have resumed their thrust westward across the Indian Ocean, to occupy Ceylon, or the drive southward, to take Fiji and Nouméa in New Caledonia, severing US links with Australia. They could strike northeastward, to occupy the Aleutian archipelago and threaten the United States via Alaska. Or they could drive on Hawaii, by first taking its solitary outpost of Midway Island. The only vector they would no longer contemplate was the northwestern one, into Soviet Siberia. Nomonhan in the summer of 1939 had taught them a salutary lesson: to avoid a swipe from the freshly revitalized Russian bear. The new Japanese empire—or what its leaders liked to call the Greater East Asia Co-Prosperity Sphere—had a gigantic perimeter to defend, which presupposed a continuing superiority at sea. That in turn depended upon a strategy of aggressive defense; and as long as the US Navy was undefeated, that defense could not be static.

The man on whose shoulders these monumental options rested was the fifty-seven-year-old Admiral Isoroku Yamamoto, commander of the Combined Fleet, architect and executor of the raid on Pearl Harbor. Born in a simple wooden house in Nagaoka, in one of the harshest parts of Japan, Yamamoto had had a tough upbringing. He was the youngest son of an impoverished samurai family, so badly off that they farmed him off to be brought up in another household. Always cold toward him, his father, Sadayoshi, a poorly paid schoolmaster, seems to have been unusual even by contemporary Japanese standards. The boy was given the name Isoroku, which means "fifty-six," for the simple reason that his father was that age at the time. When another son died, Yamamoto's father declared that he wished it could have been Isoroku instead. Understandably, the boy left home as soon as he could to join the navy, graduating with honors from the naval academy in 1904, just in time to participate in the Russo-Japanese War. His mother sent him off to war with a specially embroidered handkerchief on which his father had inscribed: "Please fall for the Emperor and your country. Even when the flower of the warrior falls, it will still have a beautiful fragrance."

As a young lieutenant, Yamamoto had fought and been wounded—losing two fingers on his left hand—at Tsushima in 1905, as well as sustaining numerous other injuries that left permanent scars to the rest of his body. It was officially recorded that he had been wounded by a Russian shell hitting his cruiser, *Nisshin*; but the truth seems to be that it had been a "cook-off" caused by unstable Japanese explosives within the breech of one of the cruiser's own guns. He had then accompanied his admiral, Togo, to the hospital bedside of the wounded and humiliated enemy admiral, Rozhestvensky. It was an encounter that clearly left a deep impression on Yamamoto, as indeed had the extraordinary Japanese triumph at Tsushima.

As noted earlier, Admiral Togo became a national hero and Tsushima a sanctified touchstone of national policy. Yamamoto, like most Japanese of his time, was brought up venerating Togo, and felt humiliated by the fetters that he saw the West once more attempt to place upon Japan after 1905, notably in the treaties of the 1920s and 1930s, which would place limitations on the growth of the Japanese navy.

Yamamoto was an early exponent of carrier warfare, and under his influence, Japan, ignoring treaty restrictions, built up the world's most imposing and most modern carrier fleet in the 1930s, a time when most navies were still wedded to the notion of big-gun warship supremacy.* He was also a passionate gambler, never missing an opportunity to play poker, and he would open a book on almost any event, even betting a dozen bottles of beer on the sinking of the two British capital ships off Malaya in December 1941, *Repulse* and *Prince of Wales*. He once all but broke the bank at Monte Carlo, and he would joke that he would love to retire from the navy and spend the rest of his life there. Although married with a family, he enjoyed the

*The big guns on warships of twelve-inch caliber and more could wreak terrible havoc on even the most heavily armored ship. They could seldom sink it in one salvo, however, unless there was a lucky shot (as with *Bismarck* sinking HMS *Hood*). The drawback in modern warfare was that a battleship had to close within twenty miles or so for its guns to be within effective range. In contrast, the carrier-borne torpedo bomber and dive bomber—as seen at the Battle of the Coral Sea, Pearl Harbor, and in the sinking of *Repulse* and *Prince of Wales*—could strike a devastating blow at three hundred miles' distance with its torpedoes and armor-piercing bombs: the kind of warfare in which capital ships never saw each other.

adoration of a geisha mistress, Chuyiko, for most of his life; he would visit her even on the eve of shipping off to war. A small man of five foot three, Yamamoto was taciturn and modest in speech, yet every once in a while he could not resist showing off, often by perilously performing a handstand on a ship's rail, even sometimes when he was an admiral.

In the early 1920s he was sent to study at Harvard—somewhat unusual for a young Japanese man at the time—and served twice as a naval attaché in Washington. He took the opportunity to travel widely, visiting the automotive industry in Detroit and the Texas oil fields. "Because I have seen the motor industry in Detroit and the oilfields of Texas [he wrote to a friend], I know Japan has no chance if she goes to war with America, or if she starts to compete in building warships." Yet his US immersion would fail him when it came to assessing how the American character would react under the challenge of total war.

In 1932 Yamamoto was put in charge of the technical manufacture of naval aviation. One result of this collaboration was the Zero, which would prove to be the best carrier-borne fighter aircraft of any country in 1941. The following year he was given command of the First Aircraft Carrier Squadron, flying his flag in the carrier *Akagi*. Then in 1936 he became deputy minister of the navy. Although it was a position that offered little influence on policy, he celebrated the appointment by giving a party for journalists, at which he performed a handstand. He found politics in Tokyo to be overlaid with menace and violence. He tried his best to slow the march to war with the United States and Great Britain. But he became so convinced that he himself stood in danger of assassination at the hands of prowar zealots that in August 1939 he readily accepted appointment as commander in chief of the Combined Fleet in order to go to sea for his own safety.

A vigorous opponent of the aggressive army clique that had embroiled Japan in the unending war with China and had taken supreme power under General Hideki Tojo in October 1941, Yamamoto had serious misgivings about the consequences of a war with the United States. (Even before Pearl Harbor he remarked, "I shall run wild for the first six months or a year, but I have absolutely no confidence for the second and third years of the fighting.") The author of the standard Japanese biography of Yamamoto, Hiroyuki Agawa, judiciously entitled his study *The Reluctant Admiral.* Like many of those in the Japanese peace lobby in the late 1930s, Yamamoto seems to have been swept along by the tide of events.

As war became inevitable in 1941, he now put all his energies and talents into planning it as vigorously as possible. For him there was no question of resignation; the samurai code demanded his obedience. In May 1939, when he feared his assassination, Yamamoto had drafted his will: "For a warrior, dying for one's country is the most honourable thing. It makes no difference whether the death is on the battlefield or at home. It is easy to die like a warrior on the battlefield but it is difficult to die for a principle. The grace of the Emperor is great, and the Japanese history is as long as eternity. I cannot help but think about Japan's future. My personal honour, life or death is of no concern."

Recollections of Togo's triumph at Tsushima swayed the gambler in Yamamoto, an exponent of airpower over the battleship, into hoping that one decisive battle, shattering the US battle fleet at anchor in Hawaii, would neutralize the United States and permit Japan to carry out its quest for empire in the Far East. Contrary to what he might have learned from his years in the United States, Yamamoto had rashly persuaded himself by the spring of 1942 that "the American fleet was a beaten, demoralized outfit that would need to be coaxed towards its own annihilation." He could not see that the surprise attack on Pearl had ensured the implacable hatred of the United States, coupled with a grim determination to destroy the attacker. "Remember Pearl Harbor" had become a slogan that would spur Americans to extraordinary acts of courage, revealing in them unexpectedly vast reserves of energy and ingenuity. As already noted, it also ensured that the Pacific War would be fought with unbridled ferocity and with no quarter given on either side.

With the strike on Pearl Harbor having turned out at best an incomplete victory, Yamamoto proposed in March 1942 a twin-pronged operation. The objective would be to seize Midway, and at the same time lure out the residue of the US Pacific Fleet, which would be destroyed in one decisive engagement—as Togo had done to Russia in 1905. At first, the army (which would have to supply the Midway invasion forces) resisted. Then came the Doolittle Raid on Tokyo of April 18—which, in the words of one senior Japanese naval officer, "passed like a shudder over Japan." It provided a major shock to Japanese security and was a personal affront to the emperor, which gave Yamamoto just the winning hand he needed.

Symbolically, the Midway operation would be launched on May 27, Japan's national day commemorating Togo's 1905 triumph. This was but a further indication of how obsessed the Japanese mentality was with this past triumph, and of Yamamoto's determination to repeat it. The scale of the operation was monumental, and entirely hubristic in conception, incorporating virtually the whole of the Combined Fleet; far bigger than the assault on Pearl Harbor, it was described by one US historian as "Byzantine in its complexity."

Once Nagumo's bombers had softened up the defenses of Midway (which were confined to a tiny area divided between its two small islands), Rear Admiral Raizo Tanaka's Transport Group would land five thousand invasion troops. Covering them would be Vice Admiral Takeo Kurita with four heavy cruisers; and farther out still would lurk Vice Admiral Nobutake Kondo's main body, comprising four more heavy cruisers and two battleships.*

*Throughout the Pacific War, the Japanese heavy cruiser was perhaps the IJN's best weapon—sturdily constructed, able to take significant punishment, packing a powerful punch, and superior to most of her US Navy counterparts.

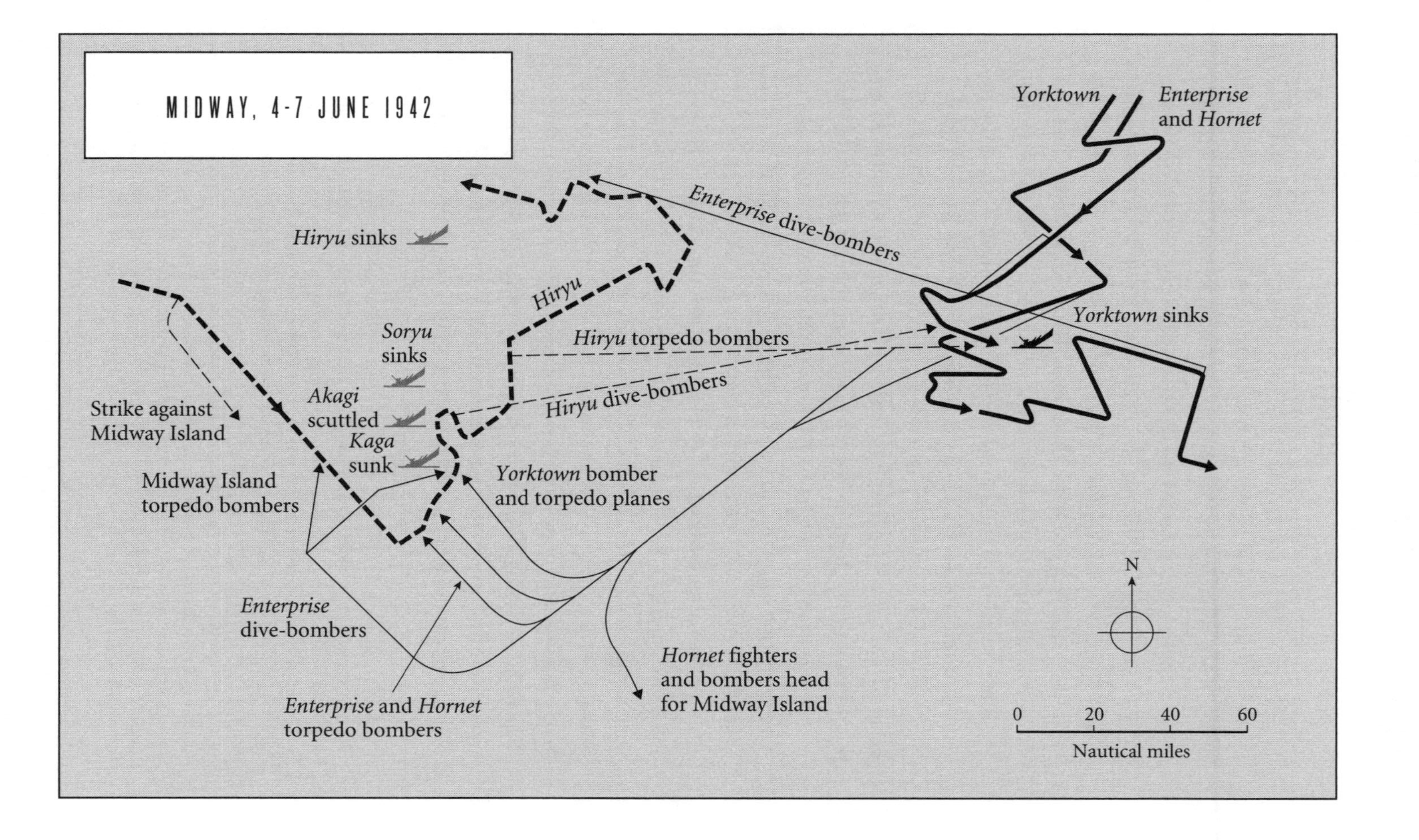
MIDWAY, 4-7 JUNE 1942
Yorktown
Enterprise and Hornet
Enterprise dive-bombers
Hiryu sinks
Hiryu
Yorktown sinks
Soryu sinks
Hiryu torpedo bombers
Hiryu dive-bombers
Akagi scuttled
Kaga sunk
Strike against Midway Island
Midway Island torpedo bombers
Yorktown bomber and torpedo planes
Enterprise dive-bombers
Enterprise and Hornet torpedo bombers
Hornet fighters and bombers head for Midway Island
N
0
20
40
60
Nautical miles

Behind that would come, some six hundred miles from Midway, at least a day's sailing apart, Yamamoto's main body, with no fewer than five battleships. He himself would command the battle from the colossal *Yamato*—at 72,800 tons, the biggest battleship ever built, whose 18.1-inch guns could outrange any existing Allied ship, and sink them with shells that weighed more than three thousand pounds each, at an extreme (and unheard-of) range of twenty-six miles.* This overwhelming force would trap, and in one decisive action smash—according to Tsushima principles—the US fleet after it had sailed to defend Midway.

Finally, as a completely separate operation, and in no way a diversionary action, Vice Admiral Moshiro Hosogaya's Northern Force would be sent north to invade three inhospitable islands in the US-owned Aleutian chain (the first US mainland territory to be taken). This operation alone mounted three heavy and three light cruisers, twelve destroyers, and two light carriers.

The fleet assembled to carry out these multiple tasks numbered some two hundred ships. It must have presented a remarkable spectacle as the main force slid out of a mist-bound Inland Sea. But the warships were to be spread over a vast area of North Pacific waters, and here would lie the gravest weakness of Yamamoto's elaborate plans: dispersal. Offending all the principles of concentration preached in the nineteenth century by the American admiral Alfred Thayer Mahan, the various components would be too far apart to support one another.

---

*Commissioned a week after Pearl Harbor, *Yamato* had been constructed in great secrecy and in clear breach of the existing naval limitation agreements. Her nine mighty guns fired just a few shots before she was sunk by US torpedo planes in April 1945, losing most of her crew of 2,332. Her sister ship *Musashi* had an equally inglorious fate, sunk at Leyte Gulf the previous year. Their demise defined a clear end to the reign of the battleship, or superdreadnought. The *Yamato* designs were approved by the Japanese Naval High Command; in all, five *Yamato*-class battleships were planned, over strong objections by naval aviators, who argued for the construction of aircraft carriers rather than battleships. It was fortunate for the Allies that Japan's battleship advocates won out.

Yamamoto's own main fleet would be lurking six hundred miles northwest of Midway, and over three thousand miles from home waters. The Aleutian element of the plan was to prove a particularly wasteful drain of resources. Great military brains from Napoleon to Clausewitz had stressed repeatedly: "Never have more than a single objective." In war, the simplest operations are the most successful. Yamamoto had too many objectives; he saw too many things. But then he was not destined to be a Napoleon, or even another "Nelson of the East."

On May 1, battle maneuvers were held aboard the mighty *Yamato*. But they were phony, sycophantically designed to please the commander in chief. With nine "hits" scored, both *Akagi* (once commanded by Yamamoto himself) and *Kaga* were sunk—but Chief of Staff Ugaki quickly reversed the umpire's decision and *Akagi* was "refloated." It was one more display of what the Japanese themselves identified as the "victory disease," which would distort all Japanese planning that spring.

Against such a dread armada as the *Kido Butai*, as the cutting edge of Japan's main naval strike force was known, what had the battered US Pacific Fleet to offer? First and foremost, it had its commander, the fifty-seven-year-old Chester W. Nimitz, known as CinCPac (pronounced "sink-pack"), the commander in chief of the Pacific Fleet. Growing up in Texas, a state not often associated with the sea, Nimitz enjoyed the homely pastime of pitching horseshoes. His grandfather had served as a sailor in the imperial German merchant marines, and Nimitz himself spoke German, which he used when studying marine engines in Germany before the First World War. The future commander in chief graduated seventh from a class of 114, but blotted his copybook when, as a young ensign, he ran a destroyer aground in the Philippines. He received a reprimand, but he redeemed himself a few years later when he saved a sailor from drowning. Nimitz became a leading US Navy authority on submarines, which was to serve him well from 1942 onward, but he had little experience of carrier warfare. When he assumed command of the Pacific Fleet in December 1941, Nimitz did so, appropriately enough, in a ceremony aboard the US submarine *Grayling*; normally this would have taken place on a battleship, but there were none left afloat at the time. The new commander combined, remarkably, the attributes of both boldness and caution, and listened to his advisers. Good at delegating authority, Nimitz was immune to panic; he commanded authority in the most adverse of times.

Upon his appointment, Nimitz had written to his wife, "I will be lucky to last six months. The public may demand action and results faster than I can produce." In fact, he was to shine as one of the few Allied commanders to provoke almost no criticism. In contrast with the Japanese brass, his assets were slender. Without a single battleship after Pearl Harbor, he had in hand just two undamaged aircraft carriers, *Enterprise* and *Hornet*. As we have seen, the latter had a largely green crew, and many of the aviators had never taken off from a carrier. His third available carrier, *Yorktown*, *Hornet*'s sister ship, was being repaired feverishly in dry dock in Pearl Harbor after taking a battering in the Battle of the Coral Sea.

But his most telling weapon lay in the realm of intelligence. Led by a backroom genius, Commander Joseph J. Rochefort Jr., a top-secret team known as Station Hypo had broken the Japanese naval code, JN-25. Admiral Ernest J. King, commander in chief of the United States Fleet and chief of naval operations (COMINCH-CNO) throughout the Second World War, had doubts about the navy intelligence teams, but Nimitz trusted them. From Rochefort's intercepts Nimitz learned that a massive Japanese assault was in the offing; he knew its components, and the rough date, but not its objective. That appeared in the Japanese code as "AF." Nimitz and Rochefort suspected that it stood for Midway, but they were not sure. So, in a clever ruse, the Americans sent an unencrpyted signal, reporting a shortage of fresh water on the island. Instantly the Japanese fell into the trap, reporting in their compromised code that "AF" was low on water.

One important additional feature was that the Japanese ships lacked the radar with which US vessels were equipped. From Rochefort's code-breaking, Nimitz could predict attacks scheduled for May 28. He assessed that the "Japanese conquest of Midway would give their bow-shaped homeland a steel tipped arrow aimed straight at the heart of Hawaii." The intelligence estimate was to prove only five minutes, five degrees, and five miles off. Nimitz's subordinate at Midway, Rear Admiral Raymond A. Spruance, would say of him: "Not only did he accept the intelligence picture, but he acted upon it at once."

One immediate setback for Nimitz was that Vice Admiral William F. "Bull" Halsey, his pugnacious carrier chief, was hospitalized with painful psoriasis. To Nimitz's surprise, Halsey nominated as his replacement Rear Admiral Spruance, placing Vice Admiral Frank Fletcher in overall command. A "black shoe" sailor who had never commanded a carrier force before, Spruance was a cool, meticulous, and careful commander, schooled on destroyers, who hated any form of publicity.* Indeed, in many ways he stood at the opposite pole to the fiery Halsey. As things turned out at Midway, Spruance's replacement of Halsey may have been a blessing in disguise. Halsey was not nicknamed "Bull" for nothing; on occasion (and dangerously) he would be the bull in the china shop. The orders Spruance received from Nimitz included the characteristic injunction: "You will be governed by the principle of calculated risk."

With foresight, and in contradistinction to Yamamoto, who rashly placed all his carriers in one basket, Nimitz divided his three into two task forces: TF 16, under Spruance, comprising *Enterprise* and *Hornet*, and TF 17, centered on the damaged *Yorktown*, escorted by two heavy cruisers and five destroyers. In command of TF 17 was Frank Fletcher, an experienced cruiser and battleship commander who had brought *Yorktown* through the Battle of the Coral Sea earlier in May.

---

*Aviators were distinguished by being permitted to wear brown shoes.

Meanwhile, a small covering force was sent to Dutch Harbor, in the Aleutians, under the command of Rear Admiral Robert A. "Fuzzy" Theobald.

On May 15, Nimitz set his modest force in motion. Repairs on *Yorktown*, estimated to require three months, were rushed through by a workforce 1,400 strong, the shipyard gangs actually taking to sea while completing their work aboard the carrier. The great ship could move only at a reduced speed. But she could launch and recover aircraft. That was enough.

Fletcher won the first hand by managing to slip his force past the submarine cordon dispatched by Yamamoto to track US naval movements out of Pearl Harbor. Arriving too late, the Japanese literally missed the boat. So, although (thanks to Station Hypo's code breakers) Nimitz had a good idea of where the major units of the Japanese fleet were heading—namely, to a rendezvous two hundred miles northwest of Midway—Yamamoto, and, more importantly, Nagumo, had no idea where the American force was, nor how many carriers it contained. They had persuaded themselves that two US carriers had been sunk in the Coral Sea, and they had no idea that *Yorktown* was in fact sailing toward Midway.

Fletcher's and Spruance's objective was to rendezvous at a map reference code-named, optimistically, Point Luck, northeast of Midway, whence they intended to ambush Nagumo's invading fleet from its left flank.

Both Nagumo's submarine and aerial reconnaissance were sloppily carried out. Had they been otherwise, Fletcher's carriers would almost certainly have been spotted, with dire consequences. The sloppy reconnaissance was all part of that "victory disease."

The Japanese armada, the greatest ever to put to sea, had set off "majestically," its crews in high spirits. In his diary, Captain Yoshitake Miwa, the Combined Fleet's air officer, wrote of this "greatest expedition": "Everybody in the fleet, from the head down to the men, entertains not the least doubt about a victory. Whence comes such a spirit that overwhelms the enemy before the battle is fought?"

Then the armada ran into a thick North Pacific fog. With visibility less than six hundred yards, the Japanese ships' speed was reduced to twelve knots, on a dangerous zigzag course. All that many could hear was the deep blast of *Yamato*'s foghorn. For security, Nagumo issued orders that pilots would have to return to their flat-tops on their own, and could not expect any radio help.

Spirits rose again as the fleet emerged from the fog near Midway. "But where is the enemy fleet?" asked Nagumo. Meanwhile, in a Catalina PBY flying boat, at 0900 on June 3, a young ensign, John Reid, shouted, "My God, aren't those ships on the horizon? I believe we have hit the jackpot." In fact this was not Nagumo's carriers, but Tanaka's invasion fleet.

At 0700 that same day, a Japanese invasion force under Rear Admiral Kakuji Kakuta, supported by the light carriers *Ryuho* and *Jun'yo*, struck Dutch Harbor in the Aleutians, and then occupied two islands, Kiska and Attu. Although these were the first US mainland territories to be seized by the enemy, they were in fact icy and barren outposts of little strategic value. But a less well informed and judicious commander than Nimitz may well have been taken in by what seemed like a ploy, designed to distract his gaze from the main attack.

The following morning, at 0545 the curtain went up on Midway with another PBY, piloted by Lieutenant William Chase, rattling off the warning: "Many planes heading Midway. Bearing 320, distance 150." Bristling with defenses that had been urgently reinforced by Nimitz, Midway at once sent all its aircraft aloft to meet the enemy—and so as not to be caught on the ground as at Pearl Harbor. Totaling 124 planes, the US force consisted primarily of a squadron of seventeen Army Air Force B-17 Flying Fortresses and a mixed bag of some three dozen Marine Corps fighters: Wildcats, Vindicators, and Brewster Buffalos. These fighters were largely obsolete; the pilots nicknamed the Vindicators "Vibrators" or "Wind-Indicators," and the stubby little Brewster Buffalos "Flying Coffins." One Marine captain said of them damningly, "Any commander that orders pilots out for combat in an F2A-3 [Buffalo] should consider the pilot as lost before leaving the ground." On some of the planes, nonmetal antiquities from peacetime, the fabric was peeling back; others developed engine failure. Was it not a disgrace to send young men to fight and die in such ancient aircraft?

Up against these planes were some of the most experienced aviators in the business, pilots who had flown in China, in the Pearl Harbor raid, the Coral Sea, and the Indian Ocean. In addition, they flew the best carrier fighter in the world, the deadly Zero (or Mitsubishi A6M). Lethally nimble and faster than any US equivalent, with a remarkable range of 1,600 miles, the Zero would later be reckoned to have shot down more than 1,500 American planes in the course of the war. Its Achilles heel was that, designed expressly for attack, it lacked protective armor, which was to prove its undoing as later US fighters evolved.*

Bombers based on Midway made several attacks on the Japanese carrier fleet. These included six TBF Avengers from *Hornet*'s VT-8,† their crews on their first combat operation, and four B-26 Marauders armed with torpedoes. Major Lofton Henderson, killed while leading his inexperienced squadron of SBU-3 Vindicator dive bombers, was the first marine aviator to die during the battle. He was posthumously honored by the main airfield at Guadalcanal being named after him

*One of the unexpected bonuses of the Aleutian campaign was that a Zero, attempting to force-land at Akuta, ended on its nose in a tundra bog, breaking the pilot's neck but itself remaining intact. The Zero was swiftly dissected by American engineers to find its secrets and weaknesses.

†US Navy carriers were designated by their commissioning date. So, for example, VB-3 was the bomber squadron from *Yorktown*; VT-6 the torpedo planes from *Enterprise*; VT-8 the torpedo planes from *Hornet*.

in August 1942. The Japanese shrugged off the attacks with almost no losses (perhaps no more than two fighters), while destroying all but one TBF torpedo bomber and two B-26s.

Out of the 108 planes Nagumo sent against Midway in that first wave, only a third were bombers. With mistaken caution, in case he should meet the naval targets that Yamamoto was counting on, he held the rest in reserve. The result was that, although shooting the US fighters out of the sky (only two remained flyable), he failed to cause more than ephemeral damage to the island's defenses. The crucial airstrip remained virtually untouched; perhaps, in their certainty of success, the Japanese were hoping to use it themselves. At the same time, unexpectedly accurate antiaircraft fire had downed a large number of the attacking aircraft. Midway could still easily resist any landing.

At 0705 on June 4, the Japanese attack leader, Lieutenant Joichi Tomonaga, flying off of *Hiryu*, sent an unhappy Admiral Nagumo this message: "There is need for a second attack wave." At first hesitant—to Yamamoto's displeasure—Nagumo complied. As no warning of the proximity of enemy carriers had reached him and with ground targets in mind, he passed down the order to replace the reserve aircraft's torpedoes with fragmentation bombs—not an operation that could be accomplished in a few minutes.* At about the same time, Nagumo had a close brush with death when a US Marine plane, crippled by antiaircraft fire from his flagship *Akagi*, crashed into her deck, narrowly missing the tightly packed bridge where he stood. Nagumo's order stood in direct contradiction of Yamamoto's: to keep the reserve strike force armed for antiship operations. But he was encouraged by this comforting message from Yamamoto: "There is no sign that our intention has been suspected by the enemy."

*Torpedoes were of use only in attacks on naval units, whereas fragmentation bombs were effective on airfields and troops. Even in a naval attack, it was sensible to arm dive bombers with armor-piercing bombs, which exploded belowdecks, instead of with fragmentation bombs, which would cause only extensive surface damage.

With such minute-by-minute actions taking place, the point at which fortune began to switch sides may be hard to determine. But Nagumo's order at 0715 on June 4 to switch missiles may have been the moment when Yamamoto's complex plan started to unravel. For if any of Yamamoto's seven battleships and heavy cruisers had been available to hammer Midway's defenses with their massive guns, not to mention the mighty 18.1-inch shells of *Yamato*, they could surely have swiftly rendered the tiny atoll's defenses untenable. But, under Yamamoto's scheme, they were being uselessly held back in reserve hundreds of miles away, for the decisive action, the reprise of Tsushima, which would never take place. Another factor with serious implications for Japanese success was that the larger warships in Yamamoto's and Kondo's forces carried scout seaplanes, an essential reconnaissance and intelligence tool—and one that was denied to Nagumo.

At about this time, Nimitz was reporting gloomily to Admiral King that, with a "great force of about 80 ships" now converging on Midway, "most of Midway's fighters, torpedo planes and dive bombers . . . were gone." Yet for all this sacrifice of lives and planes, not a single hit had been scored. Lieutenant Colonel Walter Sweeney's squadron of B-17s now entered, making several sorties. Bombing from twenty thousand feet warships that were, with utmost skill, twisting and turning at twenty-five or thirty knots has been described by one writer as like "dropping a marble from eye height on a scampering cockroach." Again, not a hit. Nevertheless, as was the custom with strategic bombing throughout the war on all fronts, exaggerated claims were made: at least one battleship was reported as seriously damaged.

The B-17s dropped 322 bombs without scoring. But they did achieve one important thing: they delayed Nagumo's jinking, zigzagging carriers from heading straight upwind at the steady twenty-five knots that was essential for launching and collecting aircraft—minutes when a carrier was at its most vulnerable.

Meanwhile, on learning that Nagumo's carriers had been identified, Fletcher, who had been holding back from committing his planes, instantly ordered an all-out attack with everything that was available. He thought it essential to get a strike in before Nagumo's superior force could land on him. Disjointedly, *Hornet* began her launches at 0700, followed promptly by *Enterprise. Yorktown*'s planes took off ninety minutes later, at 0838. But whereas the skilled Japanese crews had taken off in fifteen minutes, the green *Hornet* teams took an hour to do so. *Hornet* was to be an unlucky ship almost throughout. In the confusion of inexperience, planes had to circle, wasting precious fuel; the first attacks thus went in piecemeal, and uncoordinated, the slow and obsolete torpedo bombers disastrously unprotected by fighter cover—and many of them carrying useless torpedoes.

For all the US intelligence advantages, Fletcher still did not know where precisely in all the Pacific vastness lay Nagumo's four carriers. Worse still, led by a self-assured Commander Stanhope C. Ring, the *Hornet* attack group followed an incorrect heading of 263 degrees rather than 240, which had been angrily, but in vain, insisted on by Lieutenant Commander John C. Waldron, in command of Torpedo Squadron 8 (VT-8, from *Hornet*). In an episode later nicknamed by historians "The Flight to Nowhere," the bombers of Ring's Air Group Eight totally missed the Japanese carriers, returning to *Hornet* with all bombs "still neatly tucked under the wings."

With insubordination worthy of a Nelson, Waldron broke formation from Ring to follow what he believed to be the correct heading. It was to cost him his life, and the lives of all but one man of his squadron. Sighting the enemy carriers, Waldron began attacking at 0920, followed by VT-6, from *Enterprise*, at 0940. Waldron's last orders were: "If there is only one plane left to make the final run in, I want that man to go in and get a hit." With dauntless courage, he was last seen going in "just as straight to the Japanese Fleet as if he had a string tied to them." Without fighter escort, all fifteen TBD Devastators of VT-8 were shot down by hordes of Zeros without being able to inflict a single torpedo hit. Waldron's brave head-on attack did, however, cause Nagumo to reverse course, out of the wind, thus delaying his ability to send up more of his planes.

From his ditched plane, Ensign George H. Gay Jr., aged twenty-five, the only survivor of VT-8, saw Waldron stand heroically in the cockpit of his downed plane as the water closed in. Until that terrible morning, Gay had never carried a torpedo in a plane, let alone taken off from a ship while carrying one. Moreover, he had "never seen it done." Now, hiding afloat under his rubber dinghy, Gay uniquely had a view for the next few hours of the whole battle from the front stalls, observing Japanese decks full of planes with fuel hoses scattered all over the place—what a target! Gay was wise to play possum, his escape remarkable: three other downed US airmen were hauled aboard Nagumo's ships, brutally interrogated, and then dumped overboard—their legs attached to five-gallon canisters filled with water.

One of the three, Ensign Osmus, was evidently forced to reveal the whole US order of battle. But the revelation that he was facing three operative enemy carriers, not just two, led Nagumo into making his second fatal error. He now ordered the frantic, exhausted deck crews to switch *back* to torpedoes. The unloaded bombs were left casually on the decks, along with exposed fuel lines.

The suicidal VT-8 attack was described by one writer as being "equivalent to a stone thrown into a group of pigeons." Off *Enterprise*, VT-6 met with a similar fate to VT-8, scoring no hits to show for its efforts, thanks in part to the abysmal performance of its (US) Mark 13 aircraft torpedoes. Most traveled feet beneath their targets; or their detonators failed to explode; or they simply fell apart in the water. It would take US Navy experts many costly months yet to trace the faults and then engineer a weapon half as efficient as the Japanese Long Lance torpedo. Many brave airmen would lose their lives in vain.

An hour after the VT-8 and VT-6 strikes, the veterans of VT-3 off *Yorktown*, Fletcher's last reserve, struck with greater efficiency. Under Lieutenant Commander John Thach, the planes deployed Zero-defeating tactics, known as the Thach Weave, which they had developed from experience in the Coral Sea. Thus, *Yorktown*'s force was the only one to arrive over target and effect a coordinated attack. At a debriefing, it was assessed that "of the forty-one Devastator torpedo planes launched from three American aircraft carriers that morning, only four made it back."

Despite these horrific losses, not a single enemy target was hit. The Japanese had shot down most of the American planes and sent the rest fleeing. A final disaster occurred on the *Hornet* when a wounded pilot triggered his guns accidentally upon landing, raking the island structure and killing several personnel—including the son of the commander in chief of the Atlantic, Admiral Royal E. Ingersoll. But the American torpedo attacks had not been totally in vain. They had set the Japanese combat air patrols off course, and they had denied the Japanese carriers the time and the ability to launch a counterattack. And Lieutenant Commander John Waldron's attack had caused Nagumo to reverse course out of the wind, thus delaying his ability to put up more of his planes.

Nagumo had still not managed to get the planes of his own strike force up onto the flight deck for launch. To do so, all he needed was a short respite. He was not going to be allowed it.

CHAPTER 13

Fortune Tilts

WITH THEIR CARRIERS outnumbered and Yamamoto's vast fleet waiting in the wings to pounce, the Americans may well have thought the battle looked lost at this point. But events were moving fast. The true miracle was about to occur. Within less than half an hour, the whole course of the Pacific War would change.

At 0745, the *Enterprise*'s eldest pilot had taken off, the thirty-nine-year-old Lieutenant Commander Wade McClusky Jr., commanding a squadron of SBD Dauntless dive bombers. He was convinced, emotionally, that there was "not a chance in the world" of his being killed: "Not me, anyone else but me."* Each plane carried one lethal armor-piercing, one-thousand-pound bomb. But, on the imprecise bearings they'd been given, McClusky's planes were running low on fuel because of the time spent hunting for the enemy. Some bombers ditched from fuel exhaustion before the attack commenced. McClusky himself had never yet dropped a bomb from an SBD.

Enter US submarine *Nautilus*, a giant vessel that had been doggedly pursuing Nagumo. Spotted by Japanese destroyer *Arashi*, *Nautilus* was depth-charged relentlessly, but unsuccessfully, and chased out of Nagumo's "box." *Arashi* then returned at full speed to the Japanese fleet, throwing up a broad, white V-shaped wake. From the air,

*Awarded the Navy Cross, McClusky would indeed survive the war, to become a rear admiral.

McClusky caught sight of the wake and followed it. *Arashi* provided a fateful signpost, an arrow, leading McClusky directly to the enemy carriers.

There, clustered vulnerably close, since Nagumo's protective box had been disrupted by the earlier plane strikes, were *Kaga*, *Akagi*, and *Soryu*—the gems of the *Kido Butai*. From 15,000 feet, throwing his bomber into a near-vertical dive, McClusky selected the 38,000-ton monster *Kaga*. He narrowly escaped collision with the thirty-one-year-old Lieutenant Richard H. Best, aiming for the same target.

They went in unchallenged, the defending Zeros having been brought to low altitude by the ill-fated torpedo planes. The sacrifice had paid off. At 1022 McClusky's planes bombed the *Kaga*, killing her captain, Jisaku Okada, and the ship's senior officers. The resulting fires detonated the eighty thousand pounds of explosives that had been left strewn across the hangar deck during Nagumo's rearming process. A series of devastating multiple fuel-air explosions spread with terrifying speed through the great ship. Lieutenant Commander John Thach later declared, "I have never seen such superb dive bombing. It looked to me like almost every bomb hit."

Swiftly correcting from his near-collision with McClusky, Richard Best moved with his two wingmen four minutes later to attack the 35,000-ton *Akagi*, Nagumo's flagship. The first bomb missed, and the second bomb landed in the water. But the last bomb pierced the flight deck and exploded in the midst of a pack of bomber planes. The *Akagi* was finished.

At nearly the same time, dive bombers of *Yorktown*'s VB-3, led by Lieutenant Commander Max Leslie, hit and fatally damaged *Soryu*.

The scenes on the stricken ships beggared description. Instantaneously, *Kaga* lost 269 mechanics on her hangar deck; 419 of *Soryu*'s were similarly immolated. Fires turned the lower decks to red heat, trapping engine-room crews so that they were roasted or scalded alive. The destruction revealed basic faults in Japanese carrier design. Their flight decks were wooden, as opposed to armor-clad.* Compared with stalwart structures like *Yorktown*, or even the lost *Lexington*, the Japanese carriers were gimcrack affairs, highly flammable and with poor fire-fighting capacity, to say nothing of poor discipline, with fuel left sloshing around from an unsecured system and the bombs and torpedoes not stacked properly away. The vessels were also lacking in flash-proof doors. Like the Zero fighter, the Japanese carriers reflected the Bushido-oriented national philosophy that everything was designed for a swift, aggressive war, the minimum being done for defense, or for the protection and well-being of crews.

On *Akagi*, the proud ship Yamamoto had once commanded, now afire and dead in the water, her propeller drive destroyed, one important witness was Commander Mitsuo Fuchida, the national hero who had led the attack on Pearl Harbor. Confined to hospital quarters by a sudden appendix operation, he contrived to break out and stagger onto the flight deck just as Best's bomb struck. Narrowly escaping the blast, he moved to the bridge where he ran into Captain Minoru Genda, one of the architects of Pearl Harbor, who remarked laconically: "*Shimatta!*" ("We goofed"). It said everything. Fuchida escaped, the last man down a rope ladder, breaking both legs. In the lifeboat with him was a broken Admiral Nagumo, transferring his suddenly shattered command to the light cruiser *Nagara*. *Akagi*, the queen of flattops, the very symbol of Japanese air power, was scuttled before dawn the following day.

Of the *Soryu*'s attackers, Max Leslie, commanding VB-3, and his wingman Lieutenant (Junior Grade) P. A. Holmberg both ditched their planes near USS *Astoria* after running out of fuel, by which time their parent carrier, *Yorktown*, was under enemy attack. Leslie, Holmberg, and their gunners were rescued by one of the escorting cruisers' whaleboats.

None of *Akagi*'s planes, assembled and reloading, had had a

*The harsh realities of naval warfare in Europe had led the Royal Navy to sheet over its flight decks with armor plate. After Midway, the United States would follow suit.

chance to take off. Had they done so, and launched a revenge attack on Fletcher's carriers, the day may well have ended differently for the United States.

Moving back into the fray, the dauntless *Nautilus* fired three torpedoes into *Kaga*. All failed. One split in two and was used as a rescue float by a floundering Japanese sailor. *Kaga* and *Soryu* both sank later that evening. Meanwhile that afternoon, the fourth of Nagumo's carriers, *Hiryu*, had been pinpointed by a scout aircraft from *Yorktown* at an estimated 110 miles' distance. Spruance promptly ordered Lieutenant W. Earl Gallaher on *Enterprise* (who had taken over from a wounded McClusky) to launch her final strike of dive bombers, plus some from *Yorktown*—a total of twenty-four, all that were still flyable. It was just after 1700 hours. Despite being assaulted by more than a dozen Zeros, the attack landed four, possibly five bombs on *Hiryu*, leaving the carrier ablaze like the others. Her forward elevator was flung up against the bridge, rendering the carrier unable to launch or receive her aircraft.

A strike from *Hornet*, launched late because of a communications error that followed the "Flight to Nowhere," concentrated on the remaining escort ships, but it failed to score any hits. The day had been an undiluted fiasco for *Hornet*, the brave but unlucky ship from which the Doolittle Raiders had struck Tokyo only weeks previously.

After futile attempts at firefighting, *Hiryu* stayed afloat throughout the night of June 4. In an elaborate ceremony, accompanied with shouts of "*Banzai!*" ("Hurrah!"), the emperor's portrait was reverently removed from the ship. Despite physical attempts to rescue him, Rear Admiral Tamon Yamaguchi, flag officer of the Second Carrier Division (containing *Hiryu* and *Soryu*), insisted on staying aboard, in ancient naval tradition. Japanese legend has it that he and *Hiryu*'s captain, Tomeo Kaku, went down while calmly admiring the moon. In Yamaguchi the Imperial Japanese Navy lost possibly its most competent, and best-loved, carrier commander. His valedictory words apparently were: "I have no words to apologize for what has happened. I only wish for a stronger Japanese Navy and revenge."

Yamaguchi, and *Hiryu*, would not go entirely unavenged.

Once *Yorktown*'s location had been revealed, down like a swarm of angry bees came the Japanese torpedo planes and bombers. To Captain Elliott Buckmaster on the bridge the bombs looked "like a small keg of nails coming down." Out of eighteen bombers attacking, seven managed to score three telling hits. Fire sucked down by the ventilator shafts killed the engine-room teams, leaving the ship limping in the water. Reporting the fierce fire, Japanese aircrews declared the *Yorktown* sunk. Follow-up flights took off to sink another US carrier that had been spotted, but it was still the same, incredibly robust *Yorktown*, though already a semi-invalid from the Coral Sea. Operating with an efficiency not seen on the attacked Japanese carriers, Buckmaster's damage-control personnel managed to extinguish the fires and get *Yorktown* back in action in less than two hours, able to make nineteen knots.

Leading one of the Japanese waves was Lieutenant Joichi Tomonaga who, full of the "victory disease," had bombed Midway earlier in the day. He was shot down by Thach.

Prudently, Fletcher transferred his flag of command to the cruiser *Astoria* at 1313; he then handed over the command of the joint task force to Spruance, his better-placed junior. A short time later, a strike of torpedo planes off *Hiryu* (a last gasp that had been dispatched before destruction overtook her) connected with deadly effect. A brilliantly executed run unleashed two torpedoes into *Yorktown*'s port side, ripping "a single giant 60 × 30 foot hole in the *Yorktown*'s starboard side." Almost stationary, she began to list to an alarming 26 degrees. At about 1500 hours, Buckmaster reluctantly gave the order to abandon ship. But still the sturdy vessel did not sink; so effectively was she patched up that the second wave's torpedo bombers mistook her for *Enterprise*, convincing the High Command that two US aircraft carriers had in fact been knocked out.

Hours later, in fact not until the afternoon of June 6, an intrepid Japanese submarine, I-168, slipped in to administer the coup de grâce. Almost all the survivors off *Yorktown* had been removed to her escorting destroyer, *Hammann*. But there remained one more tragedy for the US Navy: a torpedo aimed at *Yorktown* instead struck the destroyer, splitting *Hammann* in two. As she sank, her fused depth charges exploded, killing scores of survivors struggling in the water. Valiant *Yorktown* finally went down, but most of her crew were saved, at dawn on June 7.

The Battle of Midway was over. Upon the servicemen who had fought in it, the young ensigns, lieutenants, lieutenant commanders, and ordinary GI armorers, Nimitz would heap just praise: "I firmly believe [they] have already changed the course of the war."

Thanks to the radio silence he had imposed, the fog of war, and a good measure of Japanese "victory disease," Yamamoto in his ivory tower of the gigantic *Yamato* remained supremely out of touch with reality until the very last minutes of the battle. Parallels might be found with Hitler, immured in his Wolf's Lair the previous December. At 1915 on June 5, well after the last of his four carriers had been put out of action, Yamamoto sent an absurd Combined Fleet Signal No. 158:

1. The enemy fleet, which has practically been destroyed, is retreating to the east.
2. Combined Fleet units in the vicinity are preparing to pursue the remnants and at the same time, to occupy AF.

With a crucial time lag between his distant command and Nagumo, Yamamoto seems still to have been fantasizing about his Tsushima-style naval Armageddon. Although the battle at sea was already lost, and all his carrier planes gone, Yamamoto persisted in a halfhearted attempt on Midway, sending in heavy cruisers to shell it. In the course of the action two cruisers collided in the dark. Both were seriously damaged, *Mikuma* later being sunk by the USS *Dauntless*. At last facing the facts, at 0255 on June 5 Yamamoto ordered his immense *Kido Butai* to turn about and head for home.

In sharp contrast, Nimitz back at Pearl Harbor was reckoned to be on top of events throughout. Like Yamamoto, the victorious but prudent Spruance also decided to pull back his task forces. He had been left with only three operational torpedo bombers in the whole Pacific Fleet. For his withdrawal, he was criticized at the time; doubtless Bull Halsey would have rushed in. But, as he reported to Nimitz, "I did not feel justified in risking a night encounter with possibly superior enemy forces,* but on the other hand, I did not want to be too far away from Midway next morning. I wished to be in a position from which either to follow up retreating enemy forces or to break up a landing attack on Midway." But history would vindicate Spruance: in the words of the US Navy's official historian, "Spruance's performance was superb. . . . [He] emerged from this battle one of the greatest admirals in American naval history." He would go on to command the Central Pacific Force, and later succeed Nimitz as CinCPac.

His fleet sunk around him, Nagumo had offered to take the traditional way out, but Yamamoto declared icily, "I am the only one who must apologise to his Majesty," adding, "If anyone is to commit hara-kiri because of Midway, it is I"—but he didn't. According to the ethics of Bushido, perhaps he should have done so. In two historic battles, Pearl Harbor and then Midway, Yamamoto had led his nation to ruin, a ruin from which it would never recover. But such was the legend surrounding him that he was unsackable. Fate would catch up with him the following year, when he was intercepted in an aerial ambush and killed over Bougainville in the Solomon Islands. In July 1944, during the last stages of the Battle of Saipan, Chuichi Nagumo committed suicide—not in the traditional way, but with a pistol to the temple, dying miserably in a cave.

---

*It was a form of naval warfare in which the Japanese had so far excelled.

Spurious reports that two American carriers had been sunk had briefly raised morale in the *Kido Butai.* Meanwhile, US escort vessels were picking up its survivors. As late as June 18, an American PBY flying boat saved thirty-five valuable crew members from the sunken *Hiryu.* Having given up information, none of them wanted to return to Japan, preferring to be recorded as lost with their ship.

Despite the menace of submarines, US vessels diligently kept up the search for survivors, both of ships and of downed aircraft. Ensign George H. Gay Jr. was plucked safely from his hiding place beneath his aircraft's rubber float. Machinist Albert W. Winchell, shot down off *Enterprise* together with his gunner, Aviation Radioman Douglas M. Cossett, survived seventeen days of drifting; they were the last to be rescued.

Worse was the fate, and disgracefully so, of the brave veterans of Midway repatriated to Japan, many wounded or suffering appalling burns. Officers aboard *Yamato* refused to allow any of the wounded to be taken on, on the grounds that the sight of blood might cloud the commander in chief's judgment. So they were promptly transported, amid total secrecy, to a hospital ship under cover of darkness, and then segregated from the rest of the country, forbidden to receive mail, telephone calls, or even visits from their wives. Indignant, Mitsuo Fuchida thought it was "just like we are in an internment camp!" On recovery, most of the men were shipped off to the Solomons, many never to return home or see their loved ones again.

This was all done so that imperial Japan, shunning any sense of shame at a defeat under its rigid Bushido code, could declare in banner headlines: NAVY SCORES ANOTHER EPOCHAL VICTORY. The Japanese would not learn the true story of Midway until much later in the war, when it was already lost. Even in the worst days of late 1944, their military leaders would never admit that they had lost a single battle. A total of 3,057 Japanese sailors and airmen had died at Midway. The IJN's four most powerful carriers had been sunk: *Akagi,* with 267 dead; *Kaga*, 811; *Hiryu*, 392; *Soryu*, 711. In the last phase of the battle, the heavy cruiser *Mikuma* had been sunk, with 700 crew, and *Mogami* badly damaged. Two hundred and forty-eight planes had been destroyed, the loss of their pilots representing the whole of a year's graduating class. US losses totaled one already damaged carrier (*Yorktown*), one destroyer, and 150 aircraft; a total of 307 men were killed.

It was indeed a Tsushima, but in reverse.

Bare statistics, however, don't tell the whole story. The loss of those four top aircraft carriers in the twenty minutes at Midway permanently ruined what had been the most successful Japanese naval weapon system of the war. In no way could Japan's relatively backward industrial base ever catch up with US production. More critically, the Japanese would never recover from the loss of so many planes and trained pilots. By late 1943 there would emerge an entirely new US Navy that more than made good for the losses of 1941–42. The United States would never lose another battle in the Pacific.

With the brilliance of hindsight, one can generally see a good time to end a war that one side is losing. For instance, after the First Battle of the Marne in September 1914, Moltke knew that Germany could not win in the war. Realizing, as Yamamoto did, that the war was lost for Japan after Midway, one could argue that its aftermath would have been a good moment for Tokyo to sue for peace. In that case, the hideous doomsday weapons, in the form of "Little Boy" and "Fat Man," might have remained in their boxes at Los Alamos forever.

Using strictly mathematical reasoning, Yamamoto, with his 4:3 ratio in carriers manned by crack crews, and his extensive backup of battleships and cruisers, should have won at Midway. In fact, however, he had lost before the engagement began. Instead of concentrating his forces on the main objective, as Napoleon would have done on land, Yamamoto with his ambitiously complex plans dispersed them on trivial targets such as the strategically worthless Aleutians. The two light carriers Yamamoto dispatched there could well have tipped the balance at Midway, and once Midway had been taken or neutralized, the Aleutians would have fallen into his pocket for the asking.

But, despite his veneration in Japan, Yamamoto was no Napoleon, certainly no Nelson, and no Togo, either; he was outclassed by Nimitz all along the line. Damned by the rigidity of Japanese command structures, inflated by the insidious "victory disease," he—the only man out of his forty-thousand-strong command actually to have served at Tsushima in 1905—rashly allowed himself to be seduced by that old legend. If ever there was a case of hubris affecting the course of a battle further down the lane of history, Midway was it.

Had Yamamoto succeeded at Midway, however, would it have changed the course of the war? It would certainly have prolonged it—though, as was not known then, the atom bomb was always at the end of the tunnel (the Manhattan Project began in August 1942). The A-bomb, and the devastating conventional raids on the Japanese homeland that preceded Hiroshima in 1945, were the natural consequences of the US supremacy in the skies that had first been won at Midway. But, had Midway fallen into Japanese hands, and with Hawaii under constant menace, Americans on the West Coast would naturally have been prey to biting insecurity, which could well have manifested itself in terms of pressure to reverse the strategy of "Hitler first" and the Pacific War taking second place. Across the Atlantic, the consequences of such a reversal would have been incalculable. Hitler would have been left to consolidate his grip on Europe; Soviet Russia and even England might well have succumbed. Such a Europe would have been beyond the reach of reconquest by the United States alone. Had the war run on beyond 1945, without a shadow of a doubt the atomic bomb would have been used on Germany first.

Studies have been written describing Midway as a "miracle"—or simply a gauge of skill. Perhaps it combined both. Certainly McClusky's spotting, in the whole wide Pacific, the wake of the Japanese destroyer *Arashi*, like an arrow pointing directly to the Japanese carriers, seems little less than miraculous. Great luck was certainly involved at several points. It will be recalled that Napoleon always insisted that his generals should be lucky. But that notion of luck also incorporates a great deal of skill, and a detailed study of opportunities. Napoleon would not have thought highly of Yamamoto, the unlucky admiral who did not understand the art of concentration.

Midway was indeed a most extraordinary victory. The historian Craig L. Symonds has described it as "the most complete naval victory since Horatio Nelson's near annihilation of the Spanish and French fleets at Trafalgar in 1805, and like that battle, it had momentous strategic consequences. . . . The Japanese never again seized the strategic initiative; their only hope was to hold out long enough for the Americans to tire of the struggle."

In terms of the history of naval warfare, June 4, 1942, was a stunning vindication of the pioneers' belief in the carrier and its aircraft as the future queen of the oceans. Midway saw the eclipse of the mighty *Dreadnought* class as the capital ship of navies; both the super-battleships *Musashi* and *Yamato* would be sunk by carrier planes, having scarcely fired a shot from their gigantic guns. From Midway on, the line would run directly to Hiroshima in 1945—and beyond that to the establishment of the United States as the world's naval superpower.

PART FIVE

Korea and Dien Bien Phu, 1950–1954

CHAPTER 14

The American Caesar

AS THE PACIFIC War, which Japan had thrown itself into so rashly, waged so savagely, and lost so catastrophically, came to an end in August 1945, there were some rough edges, some unfinished business, left in parts of the Far East where the Japanese had once held sway. Rough edges, unfinished business: one such legacy was in Japan's former colony of Korea. In June 1950, egged on by Stalin, Kim Il-Sung's North Korean People's Army invaded the South. At roughly the same time, organized guerrilla warfare was taking hold in France's colonial territories of Indo-China. There the Japanese capitulation had created a vacuum into which communist guerrillas, under a chieftain called Ho Chi Minh, had moved before the restored French regime and France's armed forces, barely recovered from the war, could resume control. In both conflicts, elements of hubris would display themselves in different roles, manifested in ways most challenging to Western interests.

Following the Japanese capitulation, a tacit agreement between the victorious United States and Soviet Union left that part of Korea north of the Thirty-Eighth Parallel to communist control, under a Stalinist puppet, Kim Il-Sung, while the south in effect became an American satellite state, the Republic of Korea (ROK), controlled from Japan, with its capital at Seoul, dangerously close to the border. At the head of the ROK was Syngman Rhee, an unpleasantly authoritarian figure whose declared intention was to restore national unity by force of arms. However, his small American-trained army lacked heavy weaponry and amounted to little more than a lightly armed gendarmerie. Kim Il-Sung's Democratic People's Republic of Korea, by contrast, had inherited the armor and artillery left behind by the Soviets after their August 1945 conquest of Manchuria—that is, all that had not been passed on to Stalin's fellow communist Mao Tse-tung. Over the late 1940s a series of progressively bloody border incidents took place, disquietingly reminiscent of the events in Manchuria that had led to Nomonhan in 1939.

On June 25, 1950, the North Korean army surged across the Thirty-Eighth Parallel, led by its battle-proven Russian T-34s, carrying all before it, sowing panic, and murdering prisoners of war and political "suspects." Horrific stories reached the American press of wounded GIs (a small contingent of US supporting troops had been sent hastily from Japan) thrown onto bonfires by Kim Il-Sung's horde. Seoul fell, and the North Koreans barreled on, scattering the ROK forces before them. They employed, with resounding success, the double-envelopment strategy that Zhukov had perfected not so far away at Nomonhan in 1939. By July, the South Korean army and its courageous American allies were penned into the southeastern corner of the peninsula, their backs to the historic Straits of Tsushima. Representing only 10 percent of the area of Korea, the corner came to be known as the Pusan Perimeter.

Not for the first or last time, US intelligence had been caught asleep at the wheel. At this nadir of relations with Stalin in the Cold War, there were serious fears that Korea might be just a feint, a deception, while the Soviet army, with its huge preponderance of forces, was planning the greatly feared "Big Push" into a disarmed Western Europe. In Britain army reservists were called back to the colors.

With great courage, President Harry S. Truman, who had started work humbly as a Missouri haberdasher and had unexpectedly found himself at the White House upon the death of Roosevelt in April

1945, picked up the gauntlet. He brought the North Korean invasion before the UN Security Council. The Soviet Union had vacated its seat (it was peevishly boycotting the council on account of the UN's refusal to offer a place to communist China)—so, without the customary Soviet veto, the Security Council, on June 27, was able to pass Resolution 82, which authorized a UN force to assist South Korea.

Thus the US and its Western allies could wage a war under the banner of the United Nations, a state of affairs that would never arise again. The UN empowered the American government to appoint a commander, and General Douglas MacArthur was selected. He thus became commander in chief of the United Nations Command (UNCOM), and yet still remained supreme commander for the Allied Powers (SCAP) in occupied Japan, which he had effectively controlled since accepting its surrender in September 1945. MacArthur had rejected calls to have Hirohito arraigned for war crimes along with his generals, and in thus preserving the emperor in line with age-old Japanese customs, he had come to be widely regarded almost as a kind of deity in Japan, second only to the emperor. This veneration was paralleled with that which he received in the United States as the glorious victor of the Pacific War.

All the South Korean forces were also now placed under MacArthur's command. The *New York Times* praised his appointment: "Fate could not have chosen a man better qualified to command the unreserved confidence of the people of this country." The paper would soon be eating its words. As the ROK army retreated before the North Koreans, MacArthur received permission to commit US ground forces. Most GIs instinctively disliked Korea from the start. They saw it as "a nation of sad people—oppressed, unhappy, poor, silent, and sullen." They called these people, Northerners and Southerners alike, "gooks"—and it was not a term of endearment, let alone one of respect. Ever present was the smell of human excrement; and one roadless mountain was followed by another. A typical GI remark was, "I'll fight for my country, but I'll be damned if I can see why I'm fighting to save this hellhole." Yet the first US division sent to plug the leak in the "hellhole," Major General William F. Dean's Twenty-Fourth, of whose soldiers only 20 percent had seen action in the Second World War, fought with conspicuous valor, although it was outnumbered and poorly equipped. All it could do was to put up a fighting withdrawal, falling back to the Pusan Perimeter, that toehold in the extreme southeast of the country.* It looked like a final stand. But *retreat* was not a word in MacArthur's vocabulary.

Promoted in the victorious days of the Second World War to become one of America's five-star generals of the army, Douglas MacArthur had never been an ordinary general. A number of senior commanders in that war, such as Eisenhower and Marshall, had never seen action (Patton was a rare exception), but not one had experienced as much as MacArthur. In the fighting on the western front during the First World War, he had risen to the rank of brigadier general, had been nominated for a Medal of Honor, and had been awarded the Distinguished Service Cross twice and the Silver Star seven times.

On the retired list from 1937, MacArthur was recalled to active duty in 1941 as commander of United States Army Forces in the Far East, based in the Philippines. A series of disasters followed Pearl Harbor, starting with the destruction of his air force on December 8, 1941, and the invasion of the Philippines by the Japanese. MacArthur's forces were soon compelled to withdraw to Bataan, where they held out until May 1942. In March 1942, MacArthur, his family, and his staff left the nearby fortress of Corregidor Island in PT boats and escaped to Australia. There MacArthur became supreme commander, Southwest Pacific Area Theater. For his defense of the Philippines, he was awarded a Medal of Honor, which made him and his father, Arthur MacArthur Jr., the first father and son to be so decorated.

*The Twenty-Fourth Division suffered 3,602 dead and wounded and 2,962 captured, including Major General Dean.

After more than two years of fighting in the Pacific, where he was in command of all Allied ground forces, he fulfilled a promise to liberate the Philippines. It was MacArthur who officially accepted Japan's surrender aboard the USS *Missouri* on September 2, 1945.

During the Second World War, MacArthur was *the* outstanding war hero. With his battered service cap, film star dark glasses, rumpled khaki shirt and pants, and his homely corncob pipe, his face was always in the news; he was, in fact, the pinup of all Americans. Later, as supreme commander, Southwest Pacific Area, his renown continued as he directed the strategy of island-hopping that led to the Japanese homeland. The reticent Nimitz, equally deserving of laurels for his consistent series of brilliantly executed naval triumphs, was often put in the shade. However, MacArthur's star dimmed as his faulty strategies led to grievous casualties, especially among the indigenous population, in the reconquest of the Philippines in 1944 and 1945. Many perceived the costly campaign as an ego trip as MacArthur sought to fulfill his declaration of 1942: "I shall return." In this campaign, MacArthur was awarded his third and fourth Distinguished Service Medals.

There is no doubt that MacArthur remains one of the supreme geniuses of the Second World War. But it was a genius flawed by an overweening, indeed hubristic vanity, which led to one of the most humiliating disasters in US history. His biographer William Manchester titled his treatment of MacArthur *American Caesar.* Certainly he bestrode the Pacific like a colossus for ten years; and such was his grandeur and sheer battle experience that few brave men in his entourage would stick their heads above the parapet and say "No, Sir." He was, as one of his commanders once remarked, "senior to everyone but God."

At the time of the Korean invasion of June 1950, MacArthur was seventy, but he still maintained the vigor of a man half his age. In an attack, he would always be up front with the lead platoon, or in the prow of a landing craft. He was constantly at the battlefront, making seventeen observation flights over enemy lines. When MacArthur had arrived, virtually unescorted, to accept the Japanese surrender in August 1945, Churchill commented that "[of] all the amazing deeds of the war I regard General MacArthur's personal landing at Atsugi as the bravest of the lot."

By August 1950, the North Korean attacks on the perimeter of Pusan had run out of steam, their forces and supplies exhausted. Fresh UN troops were arriving. Already MacArthur was planning a counteroffensive; not for him a strategy of slogging back up the mountainous peninsula mile by mile. Instead, recollecting the success of his island-hopping strategy of the Pacific War, which targeted linchpins in the Japanese defensive structure while simply bypassing others, he planned a bold coup that would strike far behind the North Korean lines. Originally code-named Bluehearts, then Chromite, it would land a powerful force of US Marines and armor, totaling some 75,000 troops and backed by over 250 naval vessels and airpower, at Inchon, the port for Seoul, more than one hundred miles behind the enemy lines at Pusan. Inchon stood also on the narrow waistline of Korea, and driving across it MacArthur intended to trap the broken remains of the North Korean People's Army as it reeled back from the Pusan bridgehead, severing their lifeline from the North. It seems that when drawing up his plans, MacArthur noted the Japanese invasions at Inchon in 1894 and 1904, and saw how from there they had seized all of Korea before swiftly pursuing the enemy across the Yalu River into Manchuria. There is no evidence, however, that he studied Kodama's campaign in depth or made use of it for his own. But then, MacArthur usually preferred his own counsel.

One of the remarkable aspects of the planning of this daring enterprise, dependent as it was on total surprise, was the lack of security surrounding Chromite. Officers on MacArthur's staff talked about it widely and loosely. Had the North Koreans picked up on this, they could surely have sent the Inchon landings reeling back into the sea.

What was their intelligence, and that of the Soviets, doing? There were no Richard Sorges anymore in Japan; yet one must recall that 1950–51 represented the apogee of effectiveness of the Burgess-Maclean-Philby axis of British spies working for the Soviets. Kim Philby, the most accomplished and deadliest spy of all, was then MI6's man in Washington, whereas Guy Burgess was the second secretary in the British embassy. Donald Maclean was about to take over as the head of the American Department in the Foreign Office in London. One much-respected US general serving at the time, James "Jumpin' Jim" Gavin, recalled that "the enemy repeatedly displayed an uncanny knowledge of UN troop deployment." While preparing his book on MacArthur, William Manchester was assured by that expert on fictional espionage John le Carré that "it can be assumed that (1) anything we in our government knew about Korea would have been known at the British Embassy and (2) that officers in the Embassy of the rank of these three [Burgess, Maclean, Philby] would have known what the Embassy knew."

Was this just a lapse on the part of what was at the time the world's most effective spy system? Or could there have been a more sinister explanation? Namely, did Stalin refuse to communicate vital intelligence to Kim Il-Sung because he actually *wanted* the UN troops to invade North Korea, so that Mao would be forced to intervene? Or, more simply, that the headstrong MacArthur would be lured into a trap that couldn't be resisted? Given the Machiavellian deviousness with which Stalin had acted in August 1939, such an explanation does not seem impossible.

When MacArthur first put Chromite to the Joint Chiefs of Staff, he encountered strong opposition. Its chairman, General of the Army Omar Bradley, who had led the US armies in Europe from D-Day onward, and always spoke with the voice of caution, had recently predicted that "large-scale combined amphibious operations . . . will never occur again." This was distinctly the mood of the brass assembled to discuss Chromite. There were concerns about Inchon's daunting natural defenses: its approaches were two narrow passages, which could easily be blocked with mines. It had one of the highest tides in the world—thirty-two feet, exceeded only by the Bay of Fundy north of Maine. But MacArthur argued powerfully that the North Koreans, equally conscious of Inchon's formidable defenses, would accordingly be surprised by an invasion there.

He assailed the Pentagon with his arguments. On August 23, the Joint Chiefs flew from Washington to hold a meeting at MacArthur's headquarters in Tokyo. At the briefing, members of the staff of Vice Admiral James H. Doyle, who was in command of all naval operations during the Korean War, criticized every single technical aspect of Chromite. It was quite simply impossible, and the navy didn't want to do it.

But MacArthur responded with an address of historic eloquence. He compared his plan with that of General James Wolfe at the Battle of the Plains of Abraham in 1759, brushing aside the problems of tides, hydrography, and terrain. Like the defeated French commander Louis-Joseph de Montcalm, MacArthur said, "the North Koreans would regard an Inchon landing as impossible. Like Wolfe, I could take them by surprise." He argued that his plan would "cut the enemy's supply line and seal off the entire southern peninsula. By seizing Seoul I would completely paralyse the enemy's supply system—coming and going."

MacArthur ended by laying praise unsparingly on the US Navy for its triumphs in the Pacific War. Admiral Forrest Sherman, the chief of naval operations, then rose to declare that he had heard "a great voice in a great cause." That was it: Inchon, the greatest amphibious landing since D-Day in 1944, was on. None of the military potentates present that day would foresee that a victory at Inchon would make the generalissimo appear invincible and the chiefs impotent.

MacArthur, to his constant embarrassment, suffered appallingly from seasickness. Nevertheless, on September 15, 1950, he was up in the prow of the first ship of the armada as his forces landed at Inchon, on schedule. It was, he remarked, "just like Lingayen Gulf" in 1945. His appreciation proved correct—the North Koreans were taken totally by surprise. UN casualties were light—536 dead, 2,550 wounded—but 30,000 to 40,000 North Koreans were caught in the savage US naval and aerial softening-up. That other pugilistic man-of-war of the Pacific, Admiral Bull Halsey, called it "the most masterly and audacious strategic course in all history." The beachhead was consolidated two days later. But the capture of Seoul for a second time (and there would be a third, and then a fourth, before the war ended) was not well managed; in fact, it was all but disastrous. It took MacArthur two days longer to capture Seoul from his Inchon landings than it had taken General Kodama's Japanese in 1904, despite the advantages of surprise, a huge superiority of force, and 1950s standards of mobility. The US Marines leading the thrust were no longer the battle-hardened men of Tarawa and Iwo Jima, but mostly postwar rookies for whom coming up against one of the dreaded T-34s was a fearful experience. As a consequence, the escape route of the People's Army to the north, not unlike the Falaise Gap in Normandy of August 1944, was only partially closed.* The coastal road to the east remained open to the fleeing North Koreans. Although all South Korea was liberated within fifteen days, and although the remains of Kim's broken invasion force around Pusan fell back in disarray, some thirty thousand North Koreans escaped to form cadres for a new army to be retrained in the North. Their hearts were filled with a desire for vengeance.

Before Operation Chromite was even launched on September 15, Truman issued a directive to General MacArthur that—in the blinding light of twenty-first-century hindsight—comes across as one of the most disastrous ever issued by a democratically elected leader to his military

*In August 1944, General Patton's sweeping advance out of Normandy had trapped the bulk of the German army. But the net was only partially closed, so that an important element of the survivors were able to escape and fight another day.

commander. The consequences are certainly still with us today. Truman authorized MacArthur to advance beyond the Thirty-Eighth Parallel. It was a fatal door to open to the American Caesar, his head bursting with self-confidence. His wise friend and political adviser (with ambassadorial rank) William J. Sebald had been warning him of a Japanese proverb—"In the moment of victory, tighten your helmet strap." But MacArthur was unheeding. He preferred to quote that other grandiose warlord, the British field marshal Bernard Law Montgomery, who pronounced, "Generals are never given adequate directives." So it was incumbent on the recipient to interpret the edicts of the "first consul." According to William Manchester, "Washington never told MacArthur exactly what he was expected to do"—just as long as he was winning. (Here, in another context, one might detect echoes of Hitler and the German generals, limp and unquestioning after the triumphs of 1940.)

From then on president and generalissimo would no longer be singing from the same hymn sheet. On September 27, MacArthur received the top-secret National Security Council Memorandum 81/1 from Truman reminding him that operations north of the Thirty-Eighth Parallel were authorized only if "there was no entry into North Korea by major Soviet or Chinese Communist forces, no announcements of intended entry, nor a threat to counter our operations militarily" at the time.

Now how should Caesar, triumphant at the battlefront, be expected to interpret such a "directive" from distant Rome? (That same day in Moscow, at an emergency session of the Politburo, Stalin condemned the incompetence of the People's Army command and held Soviet military advisers responsible for the defeat—for which in fact he was largely responsible.) On September 29, MacArthur restored the government of the Republic of Korea under Syngman Rhee. The following day, Defense Secretary George C. Marshall sent an eyes-only message to MacArthur: "We want you to feel unhampered tactically and strategically to proceed north of the 38th parallel." (We may wonder if the eyes of Philby and Co. also saw this message.)

What of the role of Harry S. Truman in all this? Acting under the umbrella of the United Nations, did the president of the United States, and the great General Marshall, really have the authority to grant such powers to MacArthur? Of course, the United States was providing 88 percent of the 341,000 international soldiers, as well as paying most of the costs of the Korean intervention.*

Fate had certainly dealt Truman a rough hand. In 1944 he had been selected to be Roosevelt's fourth-term running mate, largely through the wheelings and dealings of US domestic politics, and not least because of the distrust of the extreme left-wing orientations of FDR's 1940 running mate, Henry A. Wallace. Failed for West Point (on account of his eyesight), and one of few twentieth-century presidents not to have a college degree, he came of humble Missouri stock, his knowledge of the outside world limited to his service as a captain of a National Guard artillery unit in France in the First World War, where he participated in the Meuse-Argonne offensive. There he gained respect for the efficiency and courage with which he commanded his artillery battery. On returning home to Kansas City he set up a haberdashery, but it crashed. He then went into local Democratic Party politics, making his reputation as an unrelenting foe of corruption and corporate greed. He was elected to the US Senate in 1934.

Sworn in as vice president on January 20, 1945, in the three remaining months of the ailing president's life, Truman was kept at arm's length. The president and the vice president met alone together only twice during those months. FDR, sick and patrician, rarely contacted Truman, even to inform him of major issues or decisions. Truman certainly received no briefing on the development of the atomic bomb, or on the growing difficulties with Stalin and Soviet Russia. He had been

*On August 29, 1950, the British Commonwealth's Twenty-Seventh Infantry Brigade arrived at Pusan to join the UN Command, which until then included only ROK and US forces; Britain's became the third-biggest force throughout the war. Among the others were France, with 1,119 men; Colombia, with 1,068; and Luxembourg, with 44.

vice president for only eighty-two days when President Roosevelt died on April 12, 1945. Now the most powerful leader on earth, knowing virtually nothing about foreign or military affairs, he told reporters, "I felt like the moon, the stars, and all the planets had fallen on me."

Truman's first awareness of the Manhattan Project's successful development of an atom bomb seems to have come at the Potsdam Conference of July 1945; there Stalin revealed, with a smirk, that he knew more than the president of the United States did. The following month, Truman, and Truman alone, would have to face up to the awful decision to explode the new weapon over Japan. A short while later (in 1949), he would be confronted by the fact that Stalin had also constructed an A-bomb.

From then on, problems would descend thick and fast, like lava from Vesuvius, on the head of the untried new president. On March 5, 1946, in his home state of Missouri, a visiting Winston Churchill (now out of power) proclaimed the reality of the "Iron Curtain" that Stalin had slammed down across Europe. Then in 1947 came Soviet threats to Turkey and Greece (Truman responded by calling forth the aid program that would bear his name). The following year provided the most serious Soviet threat to date, when the Russians blockaded the western sectors of Berlin in 1948. Powerfully backed by Britain, Truman launched a massive airlift (1948–49) to supply Berliners until the Russians backed down. Then, in 1949, the defensive military alliance that would come to be known as the North Atlantic Treaty Organization, was created under Truman's aegis. Meanwhile, as if there was not enough trouble abroad, at home Truman was beginning to be gnawed at by the phenomenon known as McCarthyism. Then in 1950 came the thunderbolt of Korea. All this occurred in little more than one term of office.

If Truman would be marked in history as the "brave little man," he should also go down as feisty, fearless—and extremely pugnacious. Typical of the latter quality was the famous letter he wrote as an outraged father to a music critic who had panned a recital given by his daughter, Margaret: "Some day I hope to meet you. When that happens you'll need a new nose, a lot of beefsteak for black eyes, and perhaps a supporter below." So, on all these accounts, the American Caesar was eventually going to find himself up against a boss of mettle.

For such a master of grand strategy, MacArthur adopted a plan of action to move into North Korea that could hardly have been more flawed, certainly as compared with the skillful campaign conceived by the Japanese generals Kodama and Kuroki back in 1904. Admittedly their objectives had been somewhat different, but not entirely. Kuroki's had been to seize a backdoor entry to Port Arthur by crossing the Yalu River near its mouth in the west, and incidentally thereby to secure the whole Korean peninsula for Japan. After capturing Seoul, he had thrust northward for the lower reaches of the Yalu, that strategic and historic river that, over its five-hundred-mile length, traditionally demarcated the frontier of Korea and Manchuria. Early European visitors to Korea had observed that the country resembled "a sea in a heavy gale" because of the many mountain ranges that crisscrossed the peninsula. Some 80 percent of North Korea is composed of mountains and uplands, bitterly cold in winter, separated by deep and narrow valleys, particularly in the north, which run at right angles to the Yalu. In the west, however, the coastal plains are wide and flat. It was to these that Kuroki's invading force had wisely stuck in 1904.

In contrast, MacArthur sent his motley forces up to the Yalu through the mountains, dividing his already stretched command to make a further seaborne landing at Wonsan on the opposite coast to Inchon. Into the rugged middle he pushed the newly reformed ROK army, still shaky from its June drubbing. Between the various components of his forces were the numerous lateral and trackless valleys, making it virtually impossible for one unit to render assistance to its neighbor; but, on the other hand, making conditions ideal for the infiltration of a largely foot-borne army from across the Yalu. Ahead of the spread-eagled UN forces stretched the highly porous border with Chinese-held Manchuria, formed largely by the Yalu River. Here was North Korea at its very widest: almost a thousand rugged miles, reaching to within a few miles of Soviet Vladivostok in the northeast, compared with the narrow waistlines of the peninsula of little more than a hundred miles at Seoul or Pyongyang (the North Korean capital) on both sides of the Thirty-Eighth Parallel. An offensive-defensive position on either of these lines might have commended itself to a less incautious commander than MacArthur. But the Caesar of the Orient was not such a one. Already he was proclaiming "Home by Christmas," and an impatient US public, still barely recovered from the exertions of four years of war, expected him to deliver.

In the first week of October 1950, just days since the liberation of Seoul, MacArthur's UN forces moved northward across that famous Thirty-Eighth Parallel. Meanwhile, ominously, on September 30, China's foreign minister Chou En-lai had warned, through the medium of the Indian ambassador to China, K. M. Panikkar, that if the UN crossed the parallel, China would "send troops to defend North Korea." Mao's troops, in their guise as a "People's Volunteer Army," now began to slip over the Yalu in their thousands. They went unobserved by US intelligence. There followed another warning, from the Indian prime minister Jawaharlal Nehru himself. But as US-Indian relations were strained at this time, neither warning was heeded.

On October 15, on the tiny Pacific atoll of Wake Island, which had heroically though vainly held out against the Japanese in December 1941, there took place one of the most extraordinary top-level meetings of the twentieth century. President Truman flew nearly seven thousand miles to meet his generalissimo and confer about the state of the war—the site selected by MacArthur in order to be close enough to *his* command base so as not to be away from *his* troops in the field for too long. In the days of short-hop piston-engine planes, the twenty-five hours' flying would have constituted an appreciable effort on the part of the sixty-six-year-old Truman. MacArthur's seemingly discourteous refusal to meet the president on continental US territory prompted much unfavorable comment from the press. The patrician and fastidious secretary of state, Dean Acheson, declined to go along, later explaining, with exquisite diplomatic style, that, "while General MacArthur had many of the attributes of a foreign sovereign . . . and was quite as difficult as any, it did not seem wise to recognize him as one." Acheson sent along his assistant, Dean Rusk, instead. Also in the entourage were Ambassador John J. Muccio, Secretary of the Army Frank Pace Jr., Ambassador at Large Philip C. Jessup, General of the Army Omar Bradley, and the ubiquitous diplomat W. Averell Harriman. With MacArthur came the commander in chief of the Pacific Fleet (CinCPac), Admiral Arthur W. Radford.

Photographs of the two leaders meeting each other are revealing. It was their first encounter. The former haberdasher arrived after his long flight nattily clad in tie and dark city suit, MacArthur as usual in open shirt and khaki slacks, the corncob pipe not far off. Was he demonstrating rank, that while he had commanded as a one-star West Pointer in the First World War, Truman had been a mere captain in the Missouri National Guard? Whatever MacArthur's motive, Truman took instant umbrage at his sloppy attire; also, instead of saluting his commander in chief, as protocol-minded General Marshall and other five-stars would have done, MacArthur shook hands, equal to equal.

Starting at 0736, and with the formal proceedings ending at 0912, Wake Island time, this historic encounter took place without any agenda. Truman seemed unusually taciturn, venturing few interventions, manifestly discomfited by MacArthur's towering imperial demeanor, which seemed designed to overawe and disarm lesser mortals. MacArthur started in by offering reassurance:

> I believe that the formal resistance will end throughout North and South Korea by Thanksgiving. There is little resistance left in South Korea—only about 15,000 men—and those we do not destroy, the winter will. . . . The North Koreans are making the same mistake they have made before. They have not deployed in depth. When the gap is closed the same thing will happen in the north as happened in the south. It is my hope to be able to withdraw the Eighth Army to Japan by Christmas . . .

He was "no longer fearful" of Chinese intervention: "We no longer stand hat in hand. The Chinese have 300,000 men in Manchuria. Of these probably not more than 100/125,000 are distributed along the Yalu River. Only 50/60,000 could be gotten across the Yalu River. They have no Air Force. Now that we have bases for our Air Force in Korea, if the Chinese tried to get down to Pyongyang there would be the greatest slaughter."

MacArthur went on to reject an Indian proposal to inject Indian and Pakistani troops along the Korean-Manchurian and Korean-Soviet frontiers, as a buffer between these countries and the US forces in Korea: "It would be indefensible from a military point of view. I am going to put South Korean troops up there. They will be a buffer. . . . The ROK troops can handle the situation."

After some exchanges about the current "puzzling" performance of America's ally France in Indo-China, the Wake Island meeting ended. Truman declared it "a most satisfactory conference." There was a not very dignified little ceremony at its close, at which Truman pinned a (fifth) Distinguished Service Medal on MacArthur. "You have set a shining example of gallantry and tenacity in defense and of audacity in attack matched by but few operations in military history," Truman declared, before clambering back into his plane for the long flight home. On October 30, MacArthur wrote to Truman, in true imperial style: "I left the Wake Island conference with a distinct sense of satisfaction that the county's interests had been well served through the better mutual understanding and exchange of views which it afforded. I hope it will result in building a strong defense against future efforts of those who seek for one reason or another (none of them worthy) to breach the understanding between us."

That was not quite how the world would come to see it. As William Manchester notes: "Nehru in New Delhi, for example, could have provided better answers than MacArthur on Wake." In fact, each of the two principals had heard what he wanted to hear, a true dialogue of the deaf. On October 19, four days after the conference ended, the US forces captured Pyongyang. That same day, the Thirteenth Army Group of the Chinese People's Volunteer Army (PVA) began pouring across the Yalu River, 250,000 strong.

Given that the CIA with all its electronic paraphernalia was unable to spot either Egypt's march on Sinai in 1973 or the outbreak of the Iraq-Iran War seven years later, US aerial reconnaissance in 1950 might be excused for not sighting the PVA units in the daytime. It was an army that moved, and supplied itself, on its feet. Its troops marched during the nighttime and were well camouflaged during the day. This huge army traveled the nearly three hundred miles from Manchuria to the combat zone in nineteen days. Napoleon's Grande Armée could not have done better. The American military historian S. L. A. Marshall describes the PVA's approach march as "a phantom which cast no shadow."

Meanwhile, coordinated with his Inchon (Chromite) strategy, MacArthur had landed his hastily assembled X Corps at Wonsan and Iwon behind the North Korean lines, far up on the other—eastern—side of the peninsula. His aim was to trap, and wipe out, the remains of Kim's retreating army in a giant pincer movement, pinning them against the Yalu. This was enough for a paper at West Point, maybe, but it did not suffice given the daily changing facts of life at the front in Korea. It meant that as MacArthur's forces progressed northward toward the Yalu, they would find themselves in the widest part of Korea, and the wildest. They would spread out on the longest front in modern warfare. And MacArthur was operating with indifferent troops: the ROK soldiers that had just about been reconstituted after the summer disasters, and green US units that had never operated together.

On October 25, the PVA Thirteenth Army Group launched its First Phase Offensive, hitting the UN forces near the Sino-Korean border. The Chinese and US militaries finally met on November 1. Deep in North Korea at Unsan, thousands of bugle-blowing soldiers from the PVA Thirty-Ninth Army encircled the US Eighth Cavalry Regiment (part of the First Cavalry Division), assaulting the Americans from three sides.

The First Cavalry Division had fought under MacArthur tenaciously through the recapture of the Philippines in 1944, but after 1945 it had been stationed in Tokyo. There its members became known scornfully as "MacArthur's pets," on account of their ceremonial rather than active role. Many were in their early twenties, wanting free education in return for brief military service, under the GI Bill of Rights, and never expecting to be thrown into one of the nastiest wars of the twentieth century. They were poorly trained and totally unprepared by cushy ceremonial duties in Tokyo, so it was not surprising that they would not perform well under frontline conditions. The low quality of the "First Cav," however, could not be blamed on MacArthur; rather, it was the fault of the post-1945 US administration, which, pressured by popular opinion, had demobilized US forces recklessly and prematurely, at a time when the threat from aggressive world communism was at its apogee. After the Unsan debacle, other army and marine units disparagingly described the First Cav shoulder insignia as representing "the shield they never carried, the horse they never rode, the bridge they never crossed, the line they never held, and the yellow is the reason why." Yet the overall casualty statistics show a less unflattering picture: whereas in all the battles of the Second World War the First

Cav lost 734 killed in action and 3,311 wounded (236 died of wounds), in Korea they would amount to 3,811 killed with 2,086 wounded.

The effect on the US troops of this first PVA attack is vividly described by twenty-year-old Private Carl Simon, of G Company, Eighth Cavalry, who in the darkness heard firing, bugles, and shouting. No effective resistance was offered. "There was just mass hysteria," said Simon. "It was every man for himself. The shooting was terrific, there were Chinese shouting everywhere. I didn't know which way to go. In the end I just ran with the crowd. We just ran and ran until the bugles grew fainter."

The historian Max Hastings describes how Simon found himself "among thirty-five frightened fugitives, in the midst of Korea without a compass. The officers among them showed no urge to exercise any leadership. The group merely began to shuffle southwards. Most threw away their weapons. They walked for fourteen days, eating berries, waving their yellow scarves desperately but vainly to observation planes."

Simon and his companions represented but a small part of the disaster for the US forces. On November 1, continues Hastings, "strong [Chinese] forces had hit them with great determination, separating their units, then attacking them piecemeal. Batteries in transit on the roads, rifle companies in positions, found themselves under devastating fire from small arms, mortars and Katyusha rockets." By dawn, only six of the Eighth Cavalry's officers and two hundred enlisted men were still able to function. More than 170 were wounded, and the number of dead or missing uncounted. The Eighth soon ceased to exist as an organized force.

What Simon experienced would now run through a whole army: the "Great Bug-Out" contagion, as soldiers fled in all directions like bugs from beneath a lifted stone. The surprise attack forced UN troops back to the Ch'ongch'on River, but the Chinese PVA, somewhat surprisingly, did not pursue them. Perhaps like good tacticians they were probing the enemy front for points of weakness; certainly the skill with which the PVA conducted its later Second Phase Offensive would suggest that. A key date may have been November 21, when the Soviet spy Donald Maclean was appointed head of the British Foreign Office's American Desk. From there he would have been able to inform his masters in Moscow, for onward transmission to Peking, that the British prime minister Clement Attlee would veto any US deployment of the A-bomb in Korea, or any Caesar-like adventures into China proper. Meanwhile, from his general headquarters in Tokyo, MacArthur was telling the Pentagon that he needed no more reinforcements. He treated the annual phenomenon of the Yalu freezing over, and becoming a highway for Chinese infantrymen, with complete disregard. At the front, US units, scattered all over North Korea, celebrated a miserable Thanksgiving on air-dropped frozen turkey. So much for MacArthur's autumnal expectations.

Brought about by hubris on the part of one of its greatest commanders, the Bug-Out debacle is one of most shameful episodes in US military history. One may compare the 60 percent losses suffered by some GI units in the Battle of the Bulge; during the Bug-Out, US casualties at their highest never exceeded 25 percent. But then the US forces were no longer a band of brothers. Morale could hardly have been lower. The young, inexperienced GIs who found themselves suddenly on the receiving end in North Korea were terrified of being taken prisoner by the "Chicoms" (communist Chinese), this strange phantom army of men in padded jackets. US fears of Chicom captors were largely unjustified. For all the fearsome din of bugles, drums, and shouts, the Chinese were less cruel than the North Koreans; in captivity, about the worst POWs suffered was the harsh communist "re-education program."

In the United States, critics of the war on both sides were firing off salvos. James "Scotty" Reston in the *New York Times* argued that Chinese intervention might have been forestalled if the United States had offered to allow a UN peace commission "take over a buffer zone on the Korean side of the Chinese frontier" when Pyongyang had been captured. On the other hand, Peking would have heard other, more bellicose voices, such as Syngman Rhee's, saying, "The war cannot stop at the Yalu River," and the hard-line senator William F. Knowland of California asking, "Why not a neutral zone ten miles north of the Yalu River?" On November 24, the day after a joyless Thanksgiving at the front, MacArthur launched his "Home for Christmas" offensive toward the Yalu.

The following day the PVA struck with its Second Phase Offensive. Swiftly it overran the ROK II Corps at the Battle of the Ch'ongch'on River, and then decimated the US Second Infantry Division on the UN forces' right flank. Nevertheless, MacArthur continued to urge his field commanders forward four days after the first enemy breakthrough. But, on November 28, the US Eighth Army commander Lieutenant General Walton "Bulldog" Walker, at sixty-one a veteran of both world wars, announced that the UN offensive was at an end—apparently without orders from MacArthur. That same morning in Washington, a shocked Truman told his personal staff in the White House: "General Bradley called me at 6.15 this morning. He told me a terrible message had come from General MacArthur. . . . The Chinese have come in with both feet." This was the day, he recorded later, "when the bad news from Korea had changed from rumours of resistance into certainty of defeat."

This may well have been the day that Truman, progressively disillusioned with and irritated by MacArthur, determined to get rid of his American Caesar. MacArthur's vainglorious promises of speedy and complete victory had abruptly turned into a major catastrophe for the United States' arms, and for those of its UN allies up in the line. MacArthur made a desperate plea for more troops. The only unit available was the Eighty-Second Airborne Division. MacArthur asked to be allowed to deploy men from Chiang Kai-shek's Nationalist Army. On both scores he was refused. He was thinking of a "counter-invasion against vulnerable areas of the Chinese mainland." There was talk of unleashing an atomic response; but America's allies, notably, Britain's Attlee, would have none of it. Much as Truman may then have wished to be rid of MacArthur, this was virtually impossible. For all the disaster he had unleashed by his drive to the Yalu, he was still far too popular among the American public. And still he enjoyed the support of his fellow military officers on the Joint Chiefs of Staff.

At the beginning of December, the UN Command retreated in disorder—its two main components, Walker's Eighth Army in the west and General Edward Almond's Tenth Corps in the east, cut off from each other by the main mountain ranges of North Korea, and thus destroyed in detail by the relentless PVA. The Eighth Army's retreat was the longest and most humiliating in the US Army's history. Max Hastings records one American colonel, Paul Freeman of the Twenty-Third Infantry, observing bitterly to a subordinate: "Look around here. This is a sight that hasn't been seen for hundreds of years: the men of the whole United States Army fleeing from a battlefield, abandoning their wounded, running for their lives."

In ten days, troops of the US Eighth Army retreated, or ran, 120 miles. By November 30, the PVA Thirteenth Army Group had driven the Eighth Army from all of northwest Korea. On December 5, UN troops abandoned Pyongyang. Walker ordered his Eighth Army to retreat below the Thirty-Eighth Parallel. The Eighth Army was back across that border by mid-December. "We're bugging out" became the slogan of the day. Some observers reckoned that the debacle matched the collapse of the French military in 1940.

Yet among the displays of confusion and downright cowardice, there were actions of great courage. A US regimental combat team and the US First Marine Division were caught unprepared by a three-pronged encirclement by the PVA at the Chosin Reservoir in brutal wintry weather. Pulling back from Chosin to the port of Hamhung, the US Marines fell into a deadly ambush and had to fight desperately to break their way out to the sea. Losing much of their heavy weaponry, they suffered 4,418 battle and 7,313 nonbattle casualties, the latter from frostbite—against which, in the savage North Korean winter, MacArthur's forces had been little better equipped than Hitler's before Moscow in December 1941.

Meanwhile, one of the greatest supply dumps in Korea was destroyed as the mass evacuation from Hamhung was effected. Nearly two hundred shiploads of UN Command forces and matériel were evacuated to Pusan. Before escaping, the UN Command forces razed most of the city, notably the port facilities. They had been there little more than six weeks. In driving the US Tenth Corps headlong out of North Korea, Mao could boast a signal triumph. US airpower and technical superiority could avail nothing against the hordes of footslogging, bugle-blowing men in their amorphous padded jackets.

In Washington there was serious talk of pulling out of the whole of Korea. On December 16, 1950, President Truman declared a national emergency. UN morale hit rock bottom on December 23 when General Walker, the popular commander of the Eighth Army, was killed in an automobile accident. On January 4, 1951, South Korea's capital of Seoul, now largely destroyed by US fire in the course of the Inchon landings, fell for a third time, its wretched citizens subjected once more to the mass executions and torture of their northern neighbors. The entire UN front was ruptured. At home, Dean Acheson brooded on the prevalent "stench of spiritless defeat, of death of high hopes and broad purposes."

In war the pendulum swings, however. Walker's successor, Lieutenant General Matthew B. Ridgway, a Virginian, was one of the most outstanding Allied leaders to emerge from the Second World War. There he had led the US Eighty-Second Airborne, and he was greatly respected by fellow Americans and allies alike. On taking over from the deceased Walker, on December 27, 1950, Ridgway was told by MacArthur: "Eighth Army is yours, Matt. Do what you think best." It was an act of characteristic imperial grandeur, but a wise one. Acquiring the nickname "Old Iron Tits" as a result of his habit of wearing hand grenades attached to his chest harness, Ridgway restored the morale of the beaten UN forces with astonishing speed. Now dependent on a huge backup supply system for its vast forces, and dangerously overextended, the PVA in turn had become vulnerable to unchallenged US air strikes. Its casualties were immense, unacceptable even to Mao and to China's seemingly bottomless manpower reserves.

General Bradley described Ridgway's success in turning the tide of the Korean War as "the greatest feat of personal leadership in the history of the Army." A resuscitated Eighth Army struck north again in January 1951, inflicting heavy casualties on the Chinese and recapturing the ruins of unhappy Seoul (the fourth time it had changed hands) in March that year. There Ridgway sagely paused.

With the improved military situation, Truman saw the opportunity of proposing a negotiated peace. But, then, on March 24, MacArthur threw his bombshell into the works: he publicly called upon China to admit defeat. He was thus simultaneously challenging both the Chinese and his own superiors. Truman's proposed announcement was shelved. On April 5, Representative Joseph William Martin Jr., the Republican leader in the House of Representatives, read aloud on the floor of the House a letter from MacArthur critical of Truman's Europe-first policy and limited-war strategy in the Far East. The letter concluded with the clarion call: "We must win. There is no substitute for victory."

This was an intolerable act of insubordination by a general to his democratically elected commander in chief, and it would be unacceptable to any head of state. MacArthur described his démarche as merely a "military appraisal." But it was, in fact, an ultimatum to the enemy, "an attempt to intimidate Peking on pain of sanctions which neither the United States nor any other member of the UN was prepared to apply," in the words of William Manchester. Within the senior echelons of government in Washington, anger was universal. General Ridgway thought MacArthur had "cut the ground from under the President, enraged our allies, and put the Chinese in the position of suffering a severe loss of face if they so much as accepted a bid to negotiate." All agreed that MacArthur must go. Dean Acheson declared, "There was no doubt what General MacArthur deserved; the sole issue was the wisest way to administer it."

President Truman decided. In the back of his mind was the historic dispute between President Lincoln and his overambitious Civil War general George B. McClellan—another intolerable challenge to the executive's leadership.* On the morning of April 6, 1951, Truman called a meeting in his office with Marshall, Bradley, Acheson, and Harriman to discuss what should be done. Harriman was emphatically in favor of MacArthur's relief. Acheson personally supported relieving MacArthur, but he kept quiet. Generals Bradley and Marshall opposed it on military principles and as fellow Second World War soldiers. Marshall asked for more time to consider, while Acheson, the lawyer, statesman, and politician, warned Truman that he would be facing "the biggest fight of [his] administration."

The four military advisers met with Truman in his office again on April 9. According to his memoirs, Truman had already made up his own mind by this time; Bradley informed the president that the Joint Chiefs all agreed that MacArthur should go, and Marshall now added his own assent.

On April 11, 1951, Truman drafted an order to MacArthur:

> I deeply regret that it becomes my duty as President and Commander-in-Chief of the United States military forces to replace you as Supreme Commander, Allied Powers; Commander-

*Though there the situation was somewhat reversed, as Lincoln had wanted to attack and McClellan to hold back.

in-Chief, United Nations Command; Commander-in-Chief, Far East; and Commanding General, U.S. Army, Far East.

You will turn over your commands, effective at once, to Lt. Gen. Matthew B. Ridgway. You are authorized to have issued such orders as are necessary to complete desired travel to such place as you select.

My reasons for your replacement will be made public concurrently with the delivery to you of the foregoing order, and are contained in the next following message . . .

As sometimes occurs with earthshaking events in US history, there was an embarrassing slip-up between the decision and the actual delivery of the lethal order. It was decided that it should be properly delivered to MacArthur via the secretary of the army, Frank Pace. But Pace was away at the front in Korea. A press leak resulted in MacArthur's hearing of his sacking first via the media. This rather ensured an untidy and demeaning end to a great soldier, a humiliation that the American Caesar and his very articulate supporters would play up. It was a botched execution, which, noted William Manchester, "could scarcely have been administered more unwisely." One of MacArthur's senior officers, Colonel Paul Freeman, observed, "I thought his sacking was disgraceful. Sure, he had it coming. He should have been relieved. But it should have been done in a dignified way. He was an actor and an egoist, but he had been a very great man." MacArthur told his faithful friend William Sebald that what most hurt was to be "publicly humiliated after fifty-two years in the Army." In Tokyo, the deposed generalissimo relayed the news simply to his wife: "Jeannie, we are going home at last." He would return home a martyred hero.

In Tokyo, the departure of the revered Caesar from his Dai-Ichi headquarters was received with dismay. Prime Minister Shigeru Yoshida declared that MacArthur had been "looked upon by all our people with the profoundest veneration and affection. I have no words to convey the regret of our nation to see him leave." The press felt the same way. The newspaper *Asahi* declared, "Japan's recovery must be attributed solely to his guidance. We feel as if we have lost a kind of loving father," and the *Mainichi Shimbun* stated that: "MacArthur's dismissal is the greatest shock since the end of the war. He dealt with the Japanese people not as a conqueror but a great reformer. He was a noble political missionary."

As the fallen chieftain returned to his homeland, a terrific rainstorm lashed its countryside. It seemed symbolic. Professor Arthur Schlesinger wrote: "It is doubtful if there has ever been in this country so violent and spontaneous a discharge of political passion as that provoked by the President's dismissal of the General. . . . Certainly there has been nothing to match it since the Civil War." Unexpectedly and disastrously, through the flames fanned by McCarthy and his allies on the extreme right of the Republican Party, Truman became the most hated man in America. From having been a popular president, reelected on his own merits just two years previously, Truman saw his approval rating fall to 23 percent, lower than Richard Nixon's low of 25 percent during the Watergate scandal in 1974, and Johnson's of 28 percent at the height of the Vietnam War in 1968. In the *Chicago Tribune*, Robert A. Taft, the powerful Republican leader in the Senate, called in the most vitriolic terms for immediate impeachment proceedings against Truman: "President Truman must be impeached and convicted. His hasty and vindictive removal of General MacArthur is the culmination of a series of acts which have shown that he is unfit, morally and mentally, for his high office. The American nation has never been in greater danger. It is led by a fool who is surrounded by knaves."

Across the United States there was fierce criticism of the "bankrupt haberdasher" and his "traitorous" State Department. At public events Truman was booed. Veterans sent back their medals. Constituents flooded their representatives on Capitol Hill with intemperate telegrams such as: IMPEACH THE IMBECILE; WE WISH TO PROTEST THE LATEST OUTRAGE ON THE PART OF THE PIG IN THE WHITE HOUSE;

IMPEACH THE JUDAS IN THE WHITE HOUSE WHO SOLD US DOWN THE RIVER TO THE LEFT WINGERS AND THE UN . . . When invited to address Congress on his return to Washington that April, MacArthur made a thirty-four-minute speech that was interrupted thirty times by standing ovations. One congressman, overcome by emotion, shouted, "We heard God speak here today!"

As one senator observed, the whole country was "on a great emotional binge." The climax came with the hero's cavalcade accorded to MacArthur in New York City. Experts on such detail noted that the ticker tape was "four times Eisenhower's record for dumping of litter." There were many serious Americans who were deeply worried by the Sino-Soviet threat of communist world domination as evinced by the June 1950 attack on South Korea, and who thought that victory had been within MacArthur's grasp before the Chinese intervened. Later they would note that 60 percent of the UN losses in Korea—over eighty thousand of them Americans—actually followed MacArthur's recall. But this ignored the realities: MacArthur never had sufficient forces to drive up to, or hold a line on, the Yalu, let alone embark on any more ambitious offensive enterprise, even with the support of nuclear weapons. Nor did he have the backing of America's major allies, such as Britain, or of the other components of his UN force. It was all very well for MacArthur to trumpet that limited war "was like limited pregnancy." But, only five years in from the most costly war in its history, was the American public—when it put on its thinking cap, and put its emotions back in the box—really prepared to face a fresh major conflict in the Far East? And what about the potential threat to an exposed and largely undefended Europe from Stalin's Red Army? In the end, few Americans would doubt that Truman, the elected president, had been obliged to strike MacArthur down.

The conflict in Korea would drag on, inconclusively, for more than another two years of grim positional warfare, more reminiscent of the First World War than the early fluid moves in the campaign.

Repelling several Chinese offensives on a series of well-prepared defensive lines at an immense cost to the attackers (PVA losses in the whole war were 400,000 killed and 486,000 wounded), Ridgway pushed on to the Thirty-Eighth Parallel. In the latter stages of the war, amid scenes of cowardice and incompetence, there were some bright moments, such as the British Gloucester Regiment's heroic defense on the Imjin River.

At last, on July 27, 1953, three years after the war had begun, a kind of armistice, but no peace treaty, was signed. A demilitarized zone ran roughly along the line of the Thirty-Eighth Parallel, where it had all started. Calculations of total losses reach horrendous proportions: for both the Koreas, possibly four million dead out of a prewar population of some thirty million. Writers such as Jon Halliday who are generally sympathetic to the North reckon that about five hundred thousand North Korean soldiers were killed, and over two million civilians. Persistent bombing reduced the whole of the North to rubble, its industry and infrastructure totally destroyed, in scenes that made the US wartime bombing of Japan look like amateurs' work. If ever there was a case of bombing a nation "back into the stone age," as General Curtis LeMay allegedly once threatened in Vietnam, this was it; certainly nothing in Vietnam would quite match it. As for UN losses, American deaths were put at 54,246, of which roughly 20,000 were rated as "nonbattle" fatalities. The total of battle deaths among the other UN contingents came to 3,194, of which 686 were British.

Who won? In the words of Jon Halliday and Bruce Cumings, "Each side proclaims that it won, yet each actually seems to feel that it lost." They concluded that it was "an unmitigated disaster for the Korean people of North and South"—though, with the passage of time, the "disaster" looks increasingly unequal as the South emerges as one of the economic miracles of Asia, while the crazy, impoverished, pariah North of Kim Jong-un seems to slip ever further backward.

Few acts of hubris in the twentieth century were punished more savagely or more swiftly than MacArthur's, after that remarkable triumph at Inchon went so catastrophically to his head. Thrusting on to the Yalu in pursuit of total victory was a huge risk, which proved to be a frontier too far, a risk that was unjustified, the costs to world equilibrium unwarranted. Its consequences were legion, casting long shadows beyond the actual conflict of the Korean War. Korea was the first war fought by the United States that did not end in a clear-cut American victory. As far as it had involved a United Nations commitment, this proved an experience unlikely to be repeated. When it came to the Vietnam War in the 1960s, the British prime minister Harold Wilson swiftly made it clear that British troops were not going to help out this time. Rather, and similar to President Johnson in 1968, as a consequence of the unpopularity that the war, and specifically the sacking of MacArthur, had brought him, Truman declined to run for the White House again. He would be succeeded by another great Second World War warlord, General Dwight D. Eisenhower.

Henceforth, at least until Iraq and Afghanistan came along, the United States would confine itself to waging wars with limited objectives only. As the First Gulf War of 1991 would demonstrate, its leaders would pay fastidious attention to not transgressing national borders. There would be no pursuit of the Iraqi Republican Guard over the Kuwaiti border. Probably the unyielding ferocity with which the Korean War was waged led to a hardening and a prolongation of the Cold War; to a worsening of the split with China, which would not begin to heal until the Nixon-Kissinger initiative of the early 1970s; and to a consolidation of Maoism and all its attendant evils. The spectacle of a modern Western army fleeing before Mao's cotton-clad divisions was not likely to be forgotten in East Asia, no more than had been that of the destruction of tsarist forces in 1905 Manchuria. In the eyes of much of the world, it was Korea 1951 that made a great power of Mao's China.

From the Western point of view, was the war in fact worth fighting at all? "If the Korean War was a frustrating, profoundly unsatisfactory experience," writes Max Hastings in the closing words of his 1987 account, "more than thirty-five years later it still seems a struggle that the West was utterly right to fight." That surely remains true today. If the brave, raw troops of General Dean's US Twenty-Fourth Division had not been rushed to South Korea to meet Kim Il-Sung's onslaught in June 1950, who knows what apocalyptic, speculative ventures Stalin may have been tempted to try with the overwhelming power of the Red Army presence in Europe?

Could the Korean War have been fought differently? Might a more satisfactory outcome have been achieved, leading, for instance, to a peacefully reunited Korea, if MacArthur had stopped on the Thirty-Eighth Parallel? Over sixty years later, the jury is still out.

CHAPTER 15

Echoes of Verdun

"When MacArthur learned of the Korean armistice on July 27, 1953, he said, 'This is the death warrant for Indochina.' "

—WILLIAM MANCHESTER, *American Caesar*

AT THE OPPOSITE end from Korea of the great, opaque landmass of Mao's China lay the beckoning, fertile territories of Indo-China, another restive peninsula pendant from the main body of East Asia. Ever since its associated states and kingdoms came under French rule from the 1880s onward, the rich lands of *l'Indochine* had constituted the true jewel in France's imperial crown. But from the 1940s they had been wracked by war, revolt, and guerrilla activities. Following the fall of France in 1940, the colony had been administered by Vichy France. Then Japanese forces invaded, developing Indo-China as a jumping-off base for the invasion of Malaya, but leaving Vichy administrative authority nominally intact until taking full control in the last months of the war. As early as May 1941, the Viet Minh, an independence movement led by Ho Chi Minh (who had started his career as a pastry cook in Paris), opened a campaign against French and Japanese rule. In Saigon, the anticommunist State of Vietnam, led by the former emperor Bao Dai, was belatedly granted independence by France in 1949. However, by the time the long-protracted armistice had been concluded in Korea in July 1953, what would come to be known as the First Indo-

China War between Viet Minh guerrillas and the French Far East Expeditionary Corps had already been dragging on for seven years with no signs of resolution. It looked unwinnable, but also—given the modern power of the postwar French Army—unlosable. Yet within six months of the armistice, France would suffer a defeat just about as humiliating and as decisive as that inflicted upon it by the Wehrmacht in 1940, and at the hands of ill-equipped guerrillas striking out of the jungle. This would change the whole structure of realpolitik in Southeast Asia and impel the US into another role in the area, which it would not welcome. Nor would it succeed there any better than it had in Korea.

During the closing stages of the Wake Island conference on Korea (October 15, 1950), Admiral Radford, commander in chief of the Pacific Fleet, had raised another issue almost as an afterthought:

> ADMIRAL RADFORD: The situation in Indo-China is the most puzzling of all as to what we can do or what we should do . . .
> GENERAL MACARTHUR: The situation in Indo-China is puzzling. The French have 150,000 of their best troops there with an officer of the highest reputation in command. Their forces are twice what we had in the [Pusan] perimeter and they are opposed by half of what the North Koreans had. I cannot understand why they do not clean it up. They should be able to do so in four months yet we have recently seen a debacle. This brings up a question of far deeper concern. What is the capacity and calibre of the French Army? In the First World War they were excellent. In the Second World War they were poor. The present French soldier is doubtful. If the French won't fight we are up against it because the defence of Europe hinges on them. They have the flower of the French Army in Indo-China and they are not fighting. If this is so, no matter what supplies we pour in they may be of no use. The loss of territory in itself is nothing but the French failure is broader than this. I cannot understand it.

THE PRESIDENT: I cannot understand it either.

ADMIRAL RADFORD: The French seem to have no popular backing from the local Indo-Chinese. The French must train native troops. The rest of Southeast Asia—Burma, Siam—is wide open if the Chinese Communists pursue a policy of aggression. We probably have more chance of assisting in Indo-China than anywhere else. We must stiffen the backbone of the French.

MR. HARRIMAN: The French hold a key position both in Europe and Asia. I have been told by officers who were there that the French fought well in Italy. This is a matter to which we must give close thought. The French must change their attitude relative to Indo-China.

THE PRESIDENT: We have been working on the French in connection with Indo-China for years without success. . . . This is the most discouraging thing we face. Mr. Jessup and others have worked on the French tooth and nail to try and persuade them to do what the Dutch had done in Indonesia but the French have not been willing to listen.* If the French Prime Minister comes to see me, he is going to hear some very plain talk. I am going to talk cold turkey to him. If you don't want him to hear that kind of talk, you had better keep him away from me.

ADMIRAL RADFORD: Recently there were some French ships in Hawaii. I had the impression they were not anxious to go to Indo-China and were dragging their feet. They would have stayed in Pearl Harbor for six months if I had invited them . . .

This exchange speaks volumes about the new front that had already opened in the Far East, and which would eventually drag in the United States for another bloody and costly twenty-five years.

The humiliating defeat that the flower of the French Expeditionary

---

*That is, to grant independence.

Corps was to experience at the hands of "Vietnamese peasants," in the distant valley of Dien Bien Phu would shake the confidence of the Western world in its conventional forces and redraw the map of Indo-China. What lay behind this disaster of the spring of 1954 was also a kind of hubris, but of a different flavor from that which had brought defeat to MacArthur on the Yalu, Yamamoto at Midway, and Hitler before the gates of Moscow. The soldiers of France's reconstituted army certainly brought with them the still prevailing sense of racial superiority, of the European over the Asiatic, to postwar Indo-China, to which the 1941–42 defeats at the hands of the Japanese had caused shock but little modification. Yet on top of that there was inlaid something more subtle. Since 1940 the proud French Army had suffered nothing but defeat; now it sought redemption. To compensate for the humiliations of the 1940s, there were battles, campaigns, that now *had* to be won. And to win them, French commanders perforce searched back into history to latch on to victories and triumphs of the past that might by some magic of alchemy lead the way to victory in the unpromising present. One such victory would be the Battle of Verdun in 1916.

Verdun remains almost unique among the First World War battlefields as the one where the ghosts refuse to die. They have been preserved largely by the character of the battle itself, which remains the longest in any war. Its sinister fame as history's most atrocious also derives from the sheer concentration of the battlefield, where, over a period of ten months, from February to December 1916, an area smaller than Manhattan was subjected without letup to the most intensive artillery bombardment ever known. The strategy behind the German assault in February 1916 was a bizarre one; it was not to break through, or even to capture a key French city, but to lure the French Army into a trap where it would simply be "bled white" by superior German firepower. That choice of words by the German commander, General Erich von Falkenhayn, in itself suggested the image by which Verdun would forever be associated. In the event, the battle was to bleed the attacking Germans almost in equal measure. Yet well beyond World War I, Verdun would stand out in French memories as *the* amulet signifying supreme courage and supreme sacrifice.

On the handful of hills bordering Verdun on the sleepy River Meuse, just 150 miles east of Paris, the big guns of both sides exacted over eight hundred thousand French and German casualties over the ten months of battle, and so tortured the soil that whole strips of the countryside defied all attempts to return them to cultivation. Most poilus fought and died there without ever seeing the enemy. A factor peculiar to Verdun is that whereas the fighting elsewhere on the western front took place in trenches that came to be all but effaced by the passage of time, there the fighting swirled around a series of massive concrete forts that gave the city its 1914 reputation as the world's most powerful citadel. Two of them, Forts Vaux and Douaumont, scenes of some of the most lethal fighting of the battle, are kept open to the public. Battered and crumbling, they stand like Shelley's statue of Ozymandias, monuments to the folly, pride, and heroism that epitomized what we still call the Great War.

Each year visitors in their thousands troop through the rugged casemates of Douaumont; one French writer, Jean Dutourd, taken to Verdun as a boy, claims, "No historical remains I have seen since, however impressive, not even the Colosseum or the Temples of Paestum, moved me so profoundly as the Forts of Vaux and Douaumont." Nearby stands the spooky *ossuaire*, or ossuary, housing heaped in vaults visible to the casual eye the bones of one hundred thousand unidentified victims of the battle.

Over the postwar years, Verdun justly came to be regarded as France's finest hour. Through Marshal Philippe Pétain's *noria* system of rotating units into and out of the front line, some three-quarters of the 1914–18 French Army passed though *l'enfer de Verdun* and were exposed to its unforgettable horrors. Of the French Army's leaders of the 1950s, Raoul Salan and Henri Navarre both fought on the western front as teenagers in 1917 and would have been forcibly indoctrinated in the glory represented by Verdun the previous year, as well as in its pyrrhic horrors. Jean de Lattre de Tassigny, the elitist cavalryman, fighting in Salonika as a young lieutenant (he was twice wounded), had no direct association with Verdun, but he too would certainly have heard every detail from *copains* who had served there. Meanwhile, veterans' associations like Ceux de Verdun kept its memories alive in the interwar period.

At the same time, because of the horrifying losses suffered at Verdun, its symbolism also came to play a baneful role in the defeatism that brought France low in 1940. The remarkable success with which Forts Vaux and Douaumont had stood up to bombardment by Krupp's heaviest guns suggested to Secretary of War André Maginot, an *ancien combattant* who had lost his leg at Verdun, that the only way France could survive another war would be to surround itself with a chain of such forts. But this "Maginot Line" coupled a defensive strategy with defeatism. Hitler's smart young generals, such as Heinz Guderian, tackled the problem of Verdun-like static warfare with a blitzkrieg technique of fast-moving armored warfare. They went around the side and back of Maginot's fortresses, forcing them to capitulate without firing a shot.

And yet the myths, and the slogans, of Verdun, chimeric as they were, would persist in the heart of the French Army well into the 1950s. By some deep twist of fate, the French at Dien Bien Phu in 1953 would find themselves in a predicament much as Erich von Falkenhayn had devised for them in 1916; *Hell in a Very Small Place* was the title that the scholarly war correspondent Bernard B. Fall gave to his account of the battle.* Only this time it was a place of the French Army's own choosing, and the enemy's objective would be attained: to seize the French strongpoint, at whatever the cost (General

*Fall's book was published in 1966, a year before the author was killed with US forces in Vietnam.

Giap's Vietnamese troops would prove a great deal more contemptuous of losses than were Falkenhayn's Brandenburgers or Bavarians).

As previously noted, by the autumn of 1953 the war in Indo-China had been dragging on for seven years. It had been born in the vacuum created between the withdrawal of the defeated Japanese in 1945 and the return the following year of the French colonial rulers, who—in the eyes of their Eastern subjects—had suffered an even more humiliating defeat at the hands of fellow Asians less than five years previously. A comparable vacuum had not occurred in countries such as Burma and the Philippines, where British and American forces, bearing with them early prospects of independence, had expelled the Japanese. A combination of French intransigence toward decolonization, the corruptness of local rulers, and military aid from Red China had progressively imposed the stamp of orthodox communism upon what had begun, typically enough, as a nationalist, anticolonial revolt. The effective leader of the Viet Minh (League for the Independence of Vietnam) was now the sixty-three-year-old Ho Chi Minh, who claimed to have learned the meaning of revolution when in Paris in the 1920s. His military lieutenant was a forty-three-year-old doctor of political economy and a former history master, General Vo Nguyen Giap, who had waged guerrilla war against the Japanese and now employed it against the French, whom he had every reason to hate. For not only had he been imprisoned in French jails, his sister had been executed by the French, and his wife, sentenced to hard labor for life, had died in a French prison.

The war had so far cost France over 11,000 nationals killed, together with another 12,000 Foreign Legionnaires and African troops, and 14,000 Indo-Chinese. More than 120,000 were either wounded or missing. Since the politicians refused to send national servicemen to Indo-China, the cream of the French regular cadres had borne the brunt. Every three years consumed an entire class of officers from Saint-Cyr (the French national military academy, L'École Spéciale Militaire de Saint-Cyr). Six hundred billion francs were spent on the war each year—far more than the capital value of France's remaining commercial installations there. In what Charles de Gaulle dubbed "this absurd ballet," governments of the post-1945 Fourth Republic trooped on and off the stage with depressing regularity. In 1950, Georges Bidault was followed by Henri Queuille, who was followed by René Pleven; then Queuille returned, to be followed by Pleven again. Those hapless French leaders were weak men, presiding over a broken, divided, and bankrupt country—and plagued perpetually by the cancerlike presence of the most powerful communist party in Europe, which commanded roughly one-third of the seats in the French Assembly. Finally, in 1953, came Joseph Laniel, who was followed by Pierre Mendès France, the courageous Sephardic Jew who would take the unpopular but inevitable course of getting France out of Indo-China, thus ending what most Frenchmen called the *sale guerre*, or dirty war.

So eighteen successive governments had come and gone in the Fourth Republic, and none had found a way of ending the war. France's prime minister in 1953 was Joseph Laniel, a weaver from Calvados known to his associates simply as "Poor Joseph," whose ministry has been described by one French historian as "manufactured, like his speeches, with scissors and paste." In Indo-China the war was run, with little direction from Paris, by an army as eager to redeem the honor it still held to be tarnished from the capitulation of 1940 as it was to crush the Viet Minh. It had been constantly denied the men and equipment it demanded. In leadership, victory in 1945 left France with two military giants, Jacques-Philippe Leclerc and Jean de Lattre de Tassigny. Both were sent to try their hand in Indo-China. Each had the vision, and stature, to conclude negotiations with the rebel forces. They might have succeeded while France was still relatively strong vis-à-vis the Viet Minh, but, tragically for France—and maybe for the West as a whole—Leclerc was killed in 1947 in a plane crash over the Sahara. De Lattre, vainglorious and one of the most arrogant

Frenchmen ever to wear a *képi*, outrageously overbearing to subordinates, came to Hanoi claiming that his presence was worth a division of French troops (which, of course, the Fourth Republic didn't have). Nevertheless, only Marshal de Lattre, with his immense prestige as France's most distinguished Second World War soldier, could have succeeded in bullying extra resources from the overstrained French economy. He devoted these resources chiefly to the construction of a miniature Maginot Line of blockhouses guarding Hanoi and the Red River delta. He saved Hanoi in 1951, and seemed to be turning the war around; he was perhaps France's best hope in Indo-China.

Then a double tragedy struck. De Lattre had a heroic young son called Bernard, who at the age of fifteen had engineered his father's escape from the clutches of pro-German Vichy. By the end of the Second World War, Bernard had become the youngest officer ever to receive the *Médaille Militaire*, given for outstanding courage on the battlefield. In 1949 he volunteered for duty in Indo-China, where he swiftly made a name for himself in battle. In May 1951 Marshal de Lattre proudly bestowed the Croix de Guerre on his son. A few weeks later, Bernard was killed in a Viet Minh mortar attack, his body riddled with shrapnel wounds. Marshal de Lattre never recovered. The following January he died of cancer.

General Raoul Salan succeeded de Lattre as commander in chief in Indo-China. Rather than having any distinction as a field commander or strategist, Salan was a specialist in intrigue and manipulation. Nicknamed "the Mandarin" on account of his affinity for opium-puffing, under him any serious effort to regain the initiative in Indo-China went by default. With a characteristic curl of the upper lip, he dismissed Giap as a "non-commissioned officer learning to handle regiments." His officers tended not to like the Mandarin or to trust him; witness the fact that, in 1957, as soon as he was moved on to take command in Algeria, there was a mysterious attempt to assassinate him by firing a bazooka into his office, which killed one of his staff officers. Meddling in politics, Salan headed an insurrection in 1958 that resulted in the return of General de Gaulle at the head of the new French Fifth Republic. But de Gaulle had little use for schemers like Salan and brought him home from Algiers. Then, in 1961, Salan emerged again as leader of the putsch targeting de Gaulle and his endeavor to bring about a negotiated peace in Algeria.

When the putsch collapsed, Salan went into hiding. He rose again as the leader of the criminal, underground Organisation de l'Armée Secrète (OAS), which over several years committed a series of criminal terrorist attacks in both Algiers and metropolitan France. In 1962 he was caught and sentenced to death; the sentence was later commuted to life imprisonment, and he was finally granted amnesty in 1968.

In May 1953 Salan was succeeded by the three-star general Henri Navarre, the seventh and final French commander in Indo-China. With the war going badly for the French and becoming increasingly unpopular in Paris, Navarre was charged with bringing it to an honorable end. A fighting soldier who had won the Croix de Guerre in both world wars, he switched French strategy from defensive to offensive operations, forming mobile strike forces. He had no experience in the field in Indo-China. An uncommunicative figure with a fondness for cats, he declared on taking up his command: "I shall not make the same mistakes as my predecessors." His boast was to be fulfilled—certainly no other French commander erred on quite so grandiose a scale as he would.

One of Navarre's first appointments was to promote a giant brigadier, René Cogny, an artillery officer, to command the key northern sector. Taken prisoner by the Germans in 1940, Cogny escaped several times to join the French Resistance, but was recaptured and repeatedly tortured, which left him with a permanent limp. Out in Indo-China, the war in the south, around Saigon, had deteriorated into one of ambush and attrition; while in the north, the French, defensively based on de Lattre's blockhouses, had the delta fairly well under control. But it seemed to Navarre that Giap might be mounting a serious threat farther inland to the French-controlled kingdom of Laos. Otherwise the French generally held the initiative throughout Indo-China, and Giap appeared indecisive. During the summer of 1953, more French troops had become available, and tens of thousands of Vietnamese recruits were joining the French Union forces. In addition, US military aid was pouring in. On the other hand, with the end of the Korean War in sight, Red China seemed shortly free to concentrate its resources, especially American artillery pieces that had been captured in Korea, to aid the Viet Minh.

Toward the end of 1953, Navarre began to formulate a strategy aimed at both relieving the threat to Laos and inflicting a major defeat upon the Viet Minh before Chinese aid could become effective. He would lure General Giap into the kind of copybook set piece battle taught at Western staff colleges ever since Napoleon, and then crush him with superior equipment and firepower. It would also bear a curious resemblance to Falkenhayn's "bleeding white" strategy at Verdun in 1916. It seems, however, that Navarre and Cogny pursued fundamentally different strategies: Cogny envisaged a light, mobile base of operations, while Navarre was thinking in terms of a much more permanent, heavily defended fortress. By following this latter course, a modern, highly mobile army would seem to be sacrificing this mobility to lock itself away in a kind of P. C. Wren fortress (as in *Beau Geste*). As a consequence of this fundamental strategic discord, relations between the two French commanders progressively deteriorated.

On November 21, a strong French paratroop force led by three generals dropped on the isolated village of Dien Bien Phu ("the big administrative centre on the frontier") deep in Viet Minh country, close to the border of Laos, and astride Giap's main lines of communication with China. Navarre's plan was to bait the trap by establishing a powerful and aggressive garrison there. The attraction of Dien

Bien Phu lay in its possession of an airstrip (built by the Japanese) and its location in the largest valley in the area, which appeared too big to be easily sealed off. But it was fewer than one hundred miles from the Chinese border, and almost two hundred from the French bases around Hanoi; as a result, French supply aircraft would be operating at extreme range. Over the exceedingly rugged intervening terrain, land communications with Hanoi were nonexistent, so the Dien Bien Phu garrison would have to be sustained entirely by air. Aerial reconnaissance had been carried out during the dry season, and no account had been taken of the heavy rainfall at Dien Bien Phu or of the tendency for fog to fill the basin between the surrounding hills. Moreover, it seems that Navarre had been warned categorically by his air transport chief, Colonel Jean-Louis Nicot, that his service would not be in a position to maintain a permanent flow of supplies to the garrison. So, like Arnhem in the Second World War, this was to be an airborne battle. Encouragingly for the French, General Giap had no air force.

Despite these factors, and without going into the logistics of the operation too thoroughly, Navarre was confident that he could achieve his ends. He knew that Giap had artillery, but he had hitherto deployed nothing bigger than Japanese and Chinese 75 mm field guns, and he had never marshaled more than a few scattered light antiaircraft weapons with which to harass airdrop operations. Finally, Navarre's intelligence officers estimated that Giap's primitive supply system over the tortuous jungle trails would be incapable of provisioning any force attacking Dien Bien Phu with more than twenty-five thousand shells for its guns or with sufficient rice. This assessment was based on the simple rationalization that European troops could not do it. So Operation Castor now got under way.

Right from the beginning, the elliptical Navarre seems never to have made it clear to Cogny, his subordinate, that he intended Dien Bien Phu to become a fortress designed to withstand a regular siege.

Unlike the beaver, or *castor*, in French, from which the operation derived its code name, the French forces at Dien Bien Phu failed to dig themselves in with sufficient depth to withstand a siege. Cogny conceived of Dien Bien Phu more as a base for guerrilla operations against Giap's rear links, although heavy enemy concentrations in the area made it an improbable choice as an anchor for such light mobile forces. Navarre had at his disposal a total of a hundred battalions, though all but twenty-seven of these were absorbed in the static defenses of the de Lattre Line protecting the Red River delta. However, with his professional soldier's low esteem for the potential of Giap's irregulars, he further divided his forces by attempting a simultaneous large-scale operation in south Vietnam. Thus only nine battalions remained by Dien Bien Phu—whereas it would probably have required at least fifty to safeguard so large a perimeter as that delineated by the chain of hills dominating Dien Bien Phu's wide valley.

This, then, was the situation when the first paratroopers launched Operation Castor on November 21, 1953. By Christmas a series of costly failures had made it clear to the French that Dien Bien Phu could not be used as a base for offensive operations as General Cogny had hoped. The French next undertook a second battle, designed to dislodge the enemy from dangerous lookout points they had established on the surrounding heights, from which the whole valley, and its essential airstrip, could be kept under constant observation. In the first attack, misdirected French artillery fire killed fifteen of their own Algerian colonial troops. By mid-February 1954, this second battle had also been lost, and the base's garrison abandoned the crucial heights. On the seventeenth, Cogny gave the order that attacks out of Dien Bien Phu were henceforth to be limited to "light reconnaissance." The fact that two Viet Minh columns were striking simultaneously in both north and south Laos clearly demonstrated that Dien Bien Phu had already forfeited any usefulness as a base for active counterinsurgent operations. From now on its role would be purely defensive.

Meanwhile, General Giap had willingly accepted Navarre's challenge. With superhuman efforts, some fifty thousand coolies, living off a few pounds of rice a day and wheeling specially strengthened bicycles (an idea inherited from Tsuji in Malaya in 1941–42) that could carry 450-pound cargoes, were methodically moving up a siege train of over two hundred guns. Many of them were American 105 mm medium pieces, captured by the Chinese, either during the Korean Bug-Out or earlier from Chiang Kai-shek's defeated Kuomintang; these guns were heavier than anything yet used by Giap. Using the utmost ingenuity and industry, Giap's gunners burrowed through the crests of the hills wrested from the French. Thus the muzzles of the guns poked out of small tunnels, pointing directly down onto the French garrison, and presenting targets almost impossible to hit. When the defenders had been asked what would happen if the Viet Minh installed their artillery on the forward slopes facing Dien Bien Phu, the response was that this had never been done since Napoleon.

While Giap was slowly mustering his siege forces, the French could still have used their air mobility to withdraw the garrison, and, having pinned down the bulk of Giap's forces, gained a tactical victory elsewhere. But Navarre, hubristically confident in the superiority of his arms over a guerrilla force in a pitched battle, was determined to sit it out at Dien Bien Phu. His sappers, however, estimated that it would take five months to fly in the materials alone for constructing the necessary fortifications. Then it was discovered that, though it was surrounded by jungle woodlands, the actual valley of Dien Bien Phu was completely denuded of timber, essential to the construction of deep dugouts. And while Giap's men burrowed into the surrounding hills, the French did little to make Dien Bien Phu more tenable. Their reluctance to dig in, in contrast to the British and the Germans, had been noted ever since the First World War. "The trenches were often dug only to knee level," a survivor stated after the battle; "there were no shelters, no blockhouses, nothing. Even the latrines were unsandbagged." Worse still, the guns and mortars were not properly dug in and, glinting in their open firing positions, were strikingly visible to the Viet Minh up on the hills. Colonel Charles Piroth, the one-armed artillery commander at Dien Bien Phu, had refused offers of extra artillery. "I have more than I need," he declared, adding, "We know to the nearest gun what they have got. We shall win because we have some 155s with plunging fire."

When questioned by the visiting minister of defense, René Pleven, about the threat from Viet Minh artillery, another colonel retorted, "We shall fight as at Verdun." With all its pyrrhic glory, Verdun seems never to have been very far from French minds at Dien Bien Phu, and it was certainly the horrors and carnage of that battle that were to be evoked there. Nominated to command the outpost, General Christian de Castries remarked to General Cogny that the situation was "somewhat like Verdun." Yet this was only partially a valid analogy, since the French forces besieged at Dien Bien Phu had to be entirely resupplied by air on an exposed landing strip that was within range of Viet Minh artillery fire. In stark contrast, the French forces at Verdun had been resupplied by roads and railways that were completely beyond the reach of German long-range artillery. On another occasion, General Navarre told the US ambassador, Donald R. Heath, that Dien Bien Phu was "a veritable jungle Verdun which [Navarre hoped] will be attacked as it will result in terrific casualties to the Viet Minh and will not fall." To Bernard Fall, the aforementioned author of the seminal account of Dien Bien Phu, it was "Verdun transported into a tropical setting." A more recent author likened the effect of the dense shelling to that of Verdun, in that the steady rain of projectiles "ground the whole top layer of soil into fine sand and caused bunkers and trenches to implode."

Finally, at the official commission of 1955 inquiring into causes of the disaster, the question was raised, poignantly, as to whether the "fortifications . . . were as good as Verdun." To twenty-first-century readers it may seem bizarre that soldiers fighting a battle in 1954 would hark back so frequently to the circumstances of forty years previously. But there it was. Constantly repeated in that deadly valley was the brave slogan of 1916, "*Ils ne passeront pas!*" ("They shall not pass").

The fifty-one-year-old colonel Christian Marie Ferdinand de la Croix de Castries, in command of the French fortress as the battle began, was, like de Lattre, a dashing and courageous cavalryman of aristocratic lineage, trained at the elitist cavalry school at Saumur. In 1940 he had been captured by the Germans, but he escaped to rejoin the French Army and fight with the Allies through North Africa and Italy and in the South of France landings of 1944. Already wounded during the Italian campaign, when he was captured once more and made repeated escape attempts, he had been wounded again by a land mine in Indo-China that had smashed both his legs. He had then spent a year recuperating in France. Like Cogny, he walked with a cane and a limp. Always wearing the flashy fore-and-aft scarlet cap of the Spahis, he was described by contemporaries as a "swashbuckling" figure, with a passion for equestrian sports and women. Each of the outposts at Dien Bien Phu was given a female name—such as Isabelle, Gabrielle, and Huguette—reputedly after ladies in the commander's life. But was a cavalryman the ideal choice for the command of an immobile frontline fortress?

On March 13, as twenty-eight Viet Minh infantry battalions, comprising 37,500 men, braced themselves for battle, Giap opened his bombardment. Numerous explosive charges were detonated simultaneously so as to deceive Piroth's gunners as to the location of the Viet Minh guns, which, in their underground positions, proved to be almost invulnerable to both gunfire and air strikes. Terrible destruction was wreaked almost immediately on the ill-prepared French positions, and that very same day the Beatrice strongpoint fell to mass attacks by Giap's shock troops. Two days later, the Gabrielle strongpoint was captured. The fanaticism of Giap's men, who used the still-living bodies of their own casualties as duckboards to avoid being blown up on the French mines, greatly impressed, and demoralized, its Algerian defenders. The same day, Colonel Charles Piroth, who had guaranteed that no enemy cannon "would be able to fire three rounds before being destroyed," was so overcome with remorse that he blew himself up with a hand grenade. By March 31, Isabelle, consisting of a quarter of a square mile in which 1,800 men fought permanently in waist-deep swamp water, was isolated from the main garrison.

As Giap's noose drew ever tighter around de Castries' besieged camp, the politicians in Paris were wrangling over the niceties of German rearmament and the proposed European Defense Community. Foreign Minister Georges Bidault was in an advanced state of alcoholism. In London, his counterpart, Anthony Eden, was poised to rebuff any suggestion of British intervention, let alone the deployment of nuclear weapons. In the United States, Secretary of State John Foster Dulles and the Joint Chiefs of Staff talked vaingloriously about dropping a few nukes or unleashing air strikes on the Viet Minh lines gripping Dien Bien Phu. All this talk led to serious disillusionment in France, where feelings later turned to angry resentment. On April 4, "Poor Joseph" Laniel issued a desperate cry for help in the form of direct US air force intervention from carriers in the region. The State Department temporized amid much speechifying about the "domino theory," which held that if one country succumbed to communism, neighboring countries would do so too. About the only aid sent was in the form of Secretary Dulles, dispatched to an unenthusiastic London.

Meanwhile, conditions for the defenders became progressively more horrifying throughout April. They were particularly bad for the wounded, for whom there were no adequate medical supplies, hospital space, or doctors, and matters were worsened by Giap's deliberate refusal to pick up his own wounded. These problems were compounded by the heat and tropical diseases; and of course as the siege grew tighter the gravely wounded could not be evacuated. In charge of Dien Bien Phu's so-called Mobile Surgical Unit was a fifty-four-year-old army doctor, Major Paul-Henri Grauwin.* Once the airstrip was put out of use in April, Grauwin found his surgical bunker overwhelmed. The bald statistics speak for themselves: over the two months of siege, his unit registered 6,215 admissions, 739 operations, 324 evacuations, and 252 deaths; in one night he and another surgeon amputated twenty-three limbs and repaired ten abdominal and ten chest wounds. There was a saying among the German Legionnaires, "*Magen Schuss, Kopf Schuss—ist Spritzer*" ("Belly shot, head shot—then it's an overdose job")—meaning a lethal injection of morphine.

As if these appalling conditions, with dwindling supplies, were not enough, there were the all-pervasive black ants, rats that would gnaw undeterred at the wounded, and maggots that thrived on gangrenous flesh. Grauwin could never quite persuade his revolted patients that the fat white scavengers were beneficial to healing, in that they cleaned up dead tissue. By the end of the battle in May, ankle-deep in evil-smelling mud, with no electricity and 1,300 wounded, so crowded together that their limbs overlapped, Grauwin was forced to operate by candlelight.

Like an episode from a more romantic era, the ugly saga of Dien Bien Phu was modulated by the presence at Major Grauwin's right hand of the only female nurse, Lieutenant Geneviève de Galard-Terraube, who was to become lauded as "the angel of Dien Bien Phu." Lieutenant Galard, who was qualified as an in-flight nurse with the French air force, was accompanying battle casualties back to Hanoi

*Having survived the battle, and imprisonment by the Viet Minh, Grauwin wrote a searing account of his experiences, translated as *Doctor at Dien-Bien-Phu* (London, 1955).

from Dien Bien Phu on March 28 when the C-47 on which she was the *convoyeuse* was damaged upon landing, unable to take off again. At daylight the Viet Minh artillery destroyed the C-47, damaged the runway beyond repair, and left Geneviève de Galard stranded. She volunteered to work in the field hospital alongside Grauwin, her special status as the only female nurse in the camp granting her (in lieu of a bedroom) a small cell lined with parachute silk and furnished with a cot and chair. Overcoming male prejudices, she gained immense respect from the French soldiers for the selfless way in which she comforted the dying, and she was placed in charge of a forty-bed room housing some of the most gravely wounded.

Incongruously, in addition to the brave Geneviève, a number of Vietnamese and North African prostitutes were also present at Dien Bien Phu, brought there to staff the essential army *bordel.* These unfortunate women were also conscripted into helping in Major Grauwin's lazaretto.

The defenders—paratroops, legionnaires, Algerians, and Indo-Chinese detachments—continued to fight back with extraordinary courage and endurance.* By early April, Giap's men had already suffered some ten thousand casualties—representing half of the total Viet Minh forces—and their morale showed signs of ebbing as the *tricolore* still waved at Dien Bien Phu. Giap was concerned that 50 percent of all his effectives were committed to the siege. In the view of at least one American writer, "the People's Army came much closer to military failure at Dien Bien Phu than is generally believed." If only the French means, and will, had not been lacking.

Although the French at home were by and large sick to death of the war in Indo-China, those on the spot tended (as previously noted)

*It is not correct to speak only of "French" defenders. Many in fact were Foreign Legionnaires, of whom a majority were Germans escaping from the debacle of the previous war at home. In addition, there were substantial numbers of colonial Algerian *tirailleurs*, who fought with varying degrees of determination in what was for them an unpopular war.

to continue to regard Dien Bien Phu spiritually as something of a requital of honor; but if the heroism of the troops more than made up for the humiliations of 1940, the same could hardly be said of their senior commanders, some of whom, worse than the Bourbons, seem to have learned nothing and forgotten everything. Bernard Fall's remark that they (and the politicians in Paris) spent their time writing memoranda "concerned with covering themselves more against personal responsibility than against enemy attacks" has a nasty ring of the Third Republic about it. By mid-March, unappeasable hatred had grown between the cat-like Navarre and the bull-like Cogny, in a style all too reminiscent of Lords Lucan and Cardigan in the Crimean War a century earlier. Working in the same building, they communicated only by memoranda or via intermediaries, and Navarre seems never to have made his intentions properly clear to any of the executants.

Colonel de Castries, though certainly not lacking in courage, displayed a patrician distaste, as if from another age, for mud and blood. After his chief of staff had broken down and had been flown out, de Castries was not often seen outside his command post. According to Colonel Jules Roy, the author of *The Battle of Dienbienphu*, de Castries, for all his known personal courage, "timidly used his reserves to stop gaps or mend holes, not to strike. In reality he could not get them together because there were no shelters large enough to contain them." A sapper, or at least an infantry commander, may indeed have been a better choice for Dien Bien Phu than a cavalryman.

The quartermaster's branch was a mess, and air supply inept—partly, it seems, because a certain national arrogance impeded the French from benefiting from US experience. On the other hand, the French air force was undoubtedly deceived by its American colleagues (as indeed the Americans would deceive themselves in the 1960s) as to the efficacy of interdiction bombing against an Asiatic enemy in Korea. Giap's guns swiftly made the vital airstrip at Dien Bien Phu unusable. Now the second of the major French miscalculations began to exert its lethal influence. From China, Giap had massed a formidable array of antiaircraft weapons, shipped in from Korea, where they were no longer needed. French transport pilots flying into his murderous flak "envelope" compared it unfavorably with wartime experiences over Germany. Losses, hard to replace from the French air force's slender resources, mounted steadily. Day by day the Viet Minh dug out trenches like tentacles toward the French strongpoints, gradually encircling them in preparation for the final assault. Although Dien Bien Phu still held out with incredible heroism, and continued to receive small paratroop reinforcements, there was now no hope of escape for the beleaguered garrison. Moreover, much of the arms and ammunition parachuted in now fell into Giap's hands.

Conceivably, more and better tactical air support could have saved Dien Bien Phu, and at least have ensured a more durable peace in Indo-China. (It is worth noting that whereas during the 167 days of siege at Dien Bien Phu the total number of air sorties was 10,400, in 1966, during the Second Vietnam War, the weekly American total alone often exceeded 25,000.) In 1954, the only source that could have provided this extra air cover was the United States, and France pleaded repeatedly for assistance. Admiral Arthur W. Radford, the US chairman of the Joint Chiefs of Staff (the same who had expressed puzzlement at the Wake Island meeting between Truman and MacArthur in 1950), was distinctly favorable, but Secretary of State John Foster Dulles was badly misinformed and until the end was unable to appreciate the issues at stake. Against a shocking background of muddle and machinations, Dulles constantly encouraged the French to fight on beyond their means, while misleading them as to the extent of direct support the United States might be expected to provide in extremis. He is reported to have talked wildly about dropping "one or two atom bombs" on Giap, and, contingent on British acquiescence, a heavy air strike from American carriers was actually planned for the end of April. But Anthony Eden, now prime minister and bent on a course of nonintervention, was determined that Indo-China was to go to the negotiating table at the Geneva Conference the following month, a stance he had made clear to the Americans all along.

On May 4, morale at Dien Bien Phu received a bitter shock when a group of Moroccan colonial troops cut a hole through the barbed wire defenses and defected. The Viet Minh were swift to prevent the hole from being plugged by the defenders, and kept it open. Two days later, Giap brought up Russian Katyusha rockets to administer the coup de grâce. As a final illusion, the embattled defenders at first took the noise of the rockets to be that of a French relief column. Typical of the persistent French heroism in the last hours was that of a tank commander who had plaster casts removed from both arms, broken earlier in the battle, so that he could man the one remaining operational tank. The last survivor of the Chaffee light tanks that had been flown in and laboriously reassembled, poignantly christened *Douaumont* after the great fort at Verdun, was knocked out at the end of April. All its crew were killed by a direct hit from a 105 mm gun.

On May 7, with men dying at their posts from the effects of fifty-five days of constant bombardment and little sleep or food, Dien Bien Phu was overwhelmed. A French barrage broke up the first assault wave, but later that night the Viet Minh detonated a mine shaft, blowing up the strongpoint known as Eliane 2. Within a few hours they had overrun what remained of de Castries' defenses. Later that same day, Giap ordered an all-out attack against the remaining French units with over 25,000 Viet Minh (their numbers bolstered by new forces emerging from the jungle) against fewer than 3,000 garrison troops. At 1700, de Castries radioed French headquarters in Hanoi and talked with Cogny.

> **DE CASTRIES:** The Viets are everywhere. The situation is very grave. The combat is confused and goes on all around. I feel the end is approaching, but we will fight to the finish.

COGNY: Of course you will fight to the end. It is out of the question to run up the white flag after your heroic resistance.

Specifically ordered by Cogny not to fly the white flag, Dien Bien Phu never surrendered. In Paris, a female deputy was heard sobbing as Laniel announced the fall of the fortress to the National Assembly, adding the pointed note: "France must remind her allies that for seven years now the Army of the French Union has unceasingly protected a particularly crucial region of Asia, and has alone defended the interests of all."

It was the ninth anniversary of VE-Day, the victory over Germany. But unsurprisingly, celebrations in France were muted.

The calvary was not yet over, however, for the French survivors of Dien Bien Phu. Allegedly out of fear of their prisoners being liberated by a French column, the Viet Minh forced the captured French into killing marches of several hundred miles. Many of the exhausted French, as well as the badly wounded, could not keep up. They were either dispatched arbitrarily by their escorts or left by the wayside to the rats and ants. Out of the original total of 16,544, and of the 10,863 counted as having survived the siege and been taken prisoner, only 3,290 were officially repatriated four months later. Over 3,000 died in the battle, possibly 400 in the last two days' desperate fighting alone. These casualties amounted to 25 percent of the total committed, a truly terrible proportion rivaling First World War figures. They also represented roughly one-tenth of all French effectives in the whole of Indo-China. Giap's army suffered 8,000 dead and 15,000 wounded. Instead of the 25,000 shells that French intelligence had assessed were all Giap's transport could bring up, 350,000 had fallen on the garrison, including 12,000 accidentally presented to the Viet Minh by inaccurate French airdrops.

The day after the fall of Dien Bien Phu, the agenda of the Geneva Conference in peaceful Switzerland turned to Indo-China. In consequence, France under Pierre Mendès France (the government of "Poor Joseph" Laniel having been swept away in the backwash) withdrew entirely from Indo-China. Thanks to the conference, partition created the communist state of North Vietnam, triumphant and strong in victory, and the weak independent fragments of South Vietnam, Cambodia, and Laos. It was the end, with brutal swiftness, of the French dream of *l'Indochine.* Six months later, the standard of revolt against France was hoisted in Algeria, many of whose troops had fought at Dien Bien Phu and had seen France humbled by an army of Asiatic peasants. Dien Bien Phu was to cost France not only Indo-China but the rest of its empire as well.

The imbalance left behind in Vietnam was to lead directly to the American intervention. Together with the Korean War, Dien Bien Phu may well have been one of the most decisive events since 1945. They both demonstrated the superiority of an underdeveloped but impassioned irregular force over a conventional army—a superiority that has still not been overcome today.

The collapse of the hubristic notions of the French foreshadowed any number of historic developments. It would inevitably bring about the involvement of the United States claiming to be a supporter of democratic regimes confronting aggressive, expansionist communism—although members of the political left might regard such support as another form of colonialism. Despite the destruction of the idealism displayed by the French Centurions, the French soldiers who had fought so bravely and resolutely at Dien Bien Phu did not regard it as a battle lost in vain.* To them it was just one round in the war against international communism. Soon, the battleground would shift to Algeria. The same French names, such as Yves Godard and Marcel Bigeard, heroes of the one-more-lost battle, would reappear.

**Centurions* is the eponymous title of an outstandingly realistic novel of the era by Jean Lartéguy.

Confusing enemies, the French paratroopers arrived in Algeria still speaking of "*les Viets*" and convinced that the Algerian uprising too was communist-inspired. They were deceived; it was in fact a display of Arab nationalism, pure and simple, of a variety with which the West would soon become acquainted. As much as in Indo-China, the Paras (Tenth Parachute Division) arrived determined, idiosyncratically, that here was a battle that could not be lost, that had to be won.

Hard as the French fought there, defeat in Algeria led to the downfall of the Fourth Republic and the advent of de Gaulle. Then came revolt against de Gaulle himself, led by disillusioned Paras, which terminated in 1961 with the highly dangerous Revolt of the Generals, headed by Raoul Salan. The abortive Suez operation of 1956 (Operation Musketeer, an Anglo-French-Israeli plan to recapture the Suez Canal after it had been nationalized by the Egyptian president Gamal Abdel Nasser), strongly supported by the French Paras, was one more factor, engendered at Dien Bien Phu, that persuaded Frenchmen across the board of the *perfidie* of their Anglo-Saxon allies, who had so markedly let them down in Indo-China. Breaking away from NATO, France progressively decided to go its own way, a course that would continue through to its resolute opposition to the Iraq War of 2003. Such is the knock-on effect of history.

Epilogue

THE FRENCH HUMILIATION—AND for all the desperate courage shown at Dien Bien Phu, it was a deep humiliation—together with the Korean War, marked roughly the end of the first half of the twentieth century. It was a half century that would leave its mark on world history as the nastiest, cruelest, and most brutal ever recorded, and it was the first half of the century of hubris. But, with the detonation of the US bombs at Hiroshima and Nagasaki and the Soviet Union's acquisition of the same technology, the name of the game changed completely. The superpowers accepted that they could not win a total victory in the style of Napoleon or Hitler. It was altogether too dangerous, if not suicidal. Instead, wars were fought with a long reach, by proxy. The US and NATO allies would never allow themselves to come close to direct conflict with the Soviets. Instead, wars were fought in Vietnam, Africa, and other distant parts of the world by lesser allies, client states, the great powers remaining in the background. The same could be said of the various Middle Eastern flare-ups. So, to a certain extent, battles would revert to the style of the eighteenth century, with limited objectives, such as America's war in Vietnam, or President Anwar Sadat's Yom Kippur assault on Israel in 1973.

But the French defeat at Dien Bien Phu did not mark the end of the conflicts of the twentieth century that were scarred by hubris. In 1967 war broke out between Israel and its Arab neighbors; this was followed by a second round with the Yom Kippur War in 1973. No cycle of war described in this book could have displayed hubris, followed by its natural consequence of peripeteia, or reversal of circumstances, better than this two-part contest in the Middle East. I have already written about this conflict elsewhere, but it is worth a brief detour here.*

By any measure, the Six Day War of June 1967 was one of history's most outstanding campaigns, even if the sunshine has subsequently been dimmed by the realization that it was, from the Israeli side, something of a preemptive, lightning strike. The Israelis, however, had made it quite plain that if Egypt's President Nasser blockaded the Straits of Tehran, which would have closed access to the sea through to Israel's port of Eilat, a red line would be crossed. On May 23 that year, the Egyptians did just that. A nation that holds a red line to be a red line, Israel attacked first with an air strike on the morning of June 5 that destroyed on the ground almost the whole of the Egyptian air force. The Israelis then set to also destroying the air forces of both Syria and Jordan. The fighting that ensued was a veritable Old Testament victory—Samson and the jawbone of an ass, or Joshua and the walls of Jericho. The Arab armies were shattered, and unfortunate Jordan found itself shorn of all its territories, and responsibilities, on the West Bank.

The resulting sudden transformation of Israel was remarkable and, in the long term, questionable. In those six days it metamorphosed from being an underdog, under constant threat of destruction, to being a power that demanded to be taken into account. Many problems surfaced with victory—one of them the reemergence of the Palestinian problem. But what was most injurious was the effect it had on the Israelis' mind-set, from top to bottom. One eminent Israeli historian, Anita Shapira, in *Israel: A History* (2014), aptly labels the period from 1967 to 1973 "The Age of Euphoria" and goes on to quote the Psalms: "We were like unto them that dream." Euphoria, or hubris?

*I previously wrote about this conflict in *Kissinger* (2009) and *But What Do You Actually Do?* (2011).

Unfortunately for Israel, the brilliant victory in 1967, decisive though it was, did not bring about the dream of peace the country longed for. I was in Israel a couple of years later and found the whole country still on a high. In Tel Aviv I had an extraordinary encounter with one of the architects of the victory, General Ezer Weizman (later president of Israel). He asked me, as a British historian, for my view of what Israel's next move might be. I said, perhaps incautiously, that I thought Israel had one year to make a good peace. This provoked an explosion from Weizman, who, launching into an astonishingly racist diatribe about the martial inadequacies of the Egyptians, declared, "After the victory we have just won we can lick any combination of Arabs that comes against us."

Weizman, through tragic personal circumstances—his son was gravely wounded in the 1973 war—later swung from being an ultra-hawk to being an ultra-dove, but in 1969 he represented a very wide cross-section of Israeli opinion, still exultant over the Six-Day War. Shapira succinctly sums up the mood I had heard expressed by General Weizman: "Complacency reigned in Israel due to the scale of the 1967 victory. . . . The military leadership considered the Egyptian Army inferior and incapable of sacrifice and perseverance." For the first time a sense almost of invulnerability took over in Israel, a nation that had felt insecure since its creation in 1948. The standard of living soared, particularly among the new rich who thrived on construction projects funded by military contracts. The disease of hubris infected the nation from top to bottom; by 1973, the High Command assessed that there would be at least a forty-eight-hour warning if war were to break out again, which would guarantee enough time to mobilize the Israeli reserves.

In fact, when Egypt and Syria attacked on Yom Kippur in October 1973, the Arabs astonished the world. In particular, the world was surprised by Egypt's attack across the Suez Canal. Israel suffered heavy casualties, notably to its vital air force, thanks to the careful deployment of Russian antiaircraft missiles. There was near panic among the Israeli generalship. Moshe Dayan, the charismatic one-eyed minister of defense, who had earlier been heard to boast, "The Americans offer us money and advice; we accept the money and reject the advice," suffered a nervous breakdown. In private he was talking about the destruction of "the Third Temple," a metaphor for the state of Israel. There was also talk of using the Jericho weapon, the atomic bomb (possession of which Israel has not publicly acknowledged). Israel came very close to suffering the ultimate penalty of peripeteia. Its armies were defeated at both ends of the country: in the Golan Heights in the north, and in the Sinai in the south. It would almost certainly have suffered a crushing defeat but for the rushing of American arms to the front, backed up by the remarkable shuttle diplomacy of the US secretary of state Henry Kissinger. Eventually, Israel scored another remarkable victory. But only just.

All in all, the two campaigns showed every ingredient of a classical disaster, a demonstration of the pitfalls that follow the development of hubris. One could enlarge on the consequences of the 1973 Yom Kippur War at length. They are still with us: the internationalization of the war in the Middle East, with the attendant danger of its spreading. One might even judge that this second defeat of the Arab armies led—as necessity in war often does—to the invention of a new secret weapon: the human suicide bomber.

In the course of these studies I have pointed out how seldom it is that successful warlords know when to stop. A related observation, and perhaps an obvious one, is how little leaders ever seem capable of learning from history. But the real lesson, I think, is how deep-seated these failings are. The Middle Eastern case study is instructive. All nations have security concerns—they wouldn't be nations if they didn't—and it is fair to say that no nation has felt less secure than Israel.

It is easy to condemn all the follies we have been looking at, but Israel's experience shows very clearly that we humans find it difficult to resist infection by hubris. Perhaps we are most vulnerable to it at the moment of our greatest success. Yet it is more insidious than that, because the infection lingers: the heroics of Verdun led to the peripeteia of Dien Bien Phu a full thirty-eight years later, and those of Tsushima to the miscalculations of the Kwantung hotheads at Nomonhan and of Yamamoto at Midway several decades later.

And it is insidious not just over time but in its reach, as can be seen by how thoroughly it pervaded Israel after 1967; it was not just the High Command, let alone just one man, who succumbed. MacArthur's spectacular downfall happened because, seduced by his record of success, his political and military superiors allowed him, even invited him, to venture beyond the Thirty-Eighth Parallel. In other words, it was not just the generalissimo's Caesarism that brought about his misjudgments. The *Führerprinzip* that allowed Hitler to create disaster before Moscow (and elsewhere) was given life not just by his own act of will but by the connivance of millions of Germans. And the mystical nationalism that proved fatal for Japan was a political and cultural phenomenon embraced throughout the Land of the Rising Sun.

If hubris is part of the human condition—deep-seated, lingering, pervasive, and potentially lethal—what can we do to avoid it? If, as these chapters have shown, it is not just our leaders who ignore history and their own experience, we might conclude that we all have something to learn.

We can go back to where we started in the prologue, to the ancient Greeks. For Aristotle, *arete* (virtue or excellence) was something that falls between two extremes, between excess on the one hand and defect on the other. It is to be judged according to the circumstances—there is no advance prescription that tells us what to do. But we can be guided by recognizing the two extremes and looking for the mean. (Turning again to the ancient Greeks, the temple of Apollo at Delphi bore an inscription translated as "nothing in excess.") Hubris is plainly always the extreme of excess, and by accepting that it lies in wait for us—whether we are a warlord or a simple citizen—we are better equipped to avoid it.

That is not to say that we should always be cautious, or that acts of daring are wrong. Far from it. What we and our leaders need to understand is that the exuberance that follows victory all too easily leads to the wrong decision. Perhaps there are other clues that hubris is creeping up on us. As I have pointed out in these pages, a racist contempt for the enemy has often been a symptom indicating that a leader and even his people are in the grip of hubris—and heading for trouble.

While completing this book, I reread that eloquent and wise writer Albert Camus. His novel *The Plague* describes a mythical pandemic that devastates the North African town of Oran, and which is generally accepted to be an allegory for war or fascism, or some other universal, unspecified evil. The plague is defeated, and Oran celebrates its victory with a firework display witnessed by the novel's hero, Dr. Rieux. The novel concludes with this passage:

> . . . as he listened to the cries of joy arising from the town, Rieux remembered that such joy is always imperilled. He knew what those jubilant crowds did not know, but could have learned from books: that the plague bacillus never dies or disappears for good; that it can lie dormant for years and years in furniture and linen-chests; that it bides its time in bedrooms, cellars, trunks and bookshelves; and that perhaps the day would come when, for the bane and the enlightening of men, it would rouse up its rats again and send them forth to die in a happy city.

With the world facing ever more menacing dangers from ambitious leaders, from gangs of warlords, and from terrorists, we should note—as the ancient Greeks did—the terrible penalties that befall those who release from Pandora's box the dormant bacillus of hubris.

ACKNOWLEDGMENTS

AFTER RUSKIN CASTIGATED Whistler for charging two hundred guineas for two days' labor, Whistler replied, crushingly, "No, I ask you for the knowledge I have gained in the work of a lifetime." Attempting to express the depth of appreciation I owe over *Hubris* somewhat echoes Whistler—at least in terms of *gestation*. Should my acknowledgments cover only the three years *Hubris* actually took to write, or should they go back further?

I was born a few years after the First World War, surrounded by its poignant memories; so most of my childhood perforce consisted of thinking about war. As a teenage schoolboy during the Second World War, I was given the task of keeping up the school's "operation map," moving small flags backward and forward. While the actual battles were raging on the plains of Russia and in the Pacific, it was all very immediate. Most of the time, the flags bearing the sinister swastika and the rising sun seemed to move remorselessly forward.

After I joined up in 1943, aged seventeen, life evolved seamlessly into active education in warfare.

I began writing my first book on war, *The Price of Glory*, in 1960. Should this date, in fact, be a marker on the road to acknowledgments? If so, I remain indebted to the "guru" of Medmenham, the late Sir Basil Liddell Hart, whose teaching may have lost some credibility over the passage of time, but who taught one to *think* about war. In succession came the "Captain Professor," Sir Michael Howard. To him I owe a continuing gratitude for his (to use his favorite word) "judicious" analysis of history. There followed in his wake the new wave of "young lions" of military historians, headed by the illustrious Sir John Keegan, who, in his seminal *The Face of Battle*, made us all look at warfare in a different dimension.

I have drawn the most enormous benefit from my association with St. Antony's College, Oxford, for over forty years. It is often unquantifiable. This particularly applies to all the help I have received with *Hubris* from the Nissan, Russian, and Middle East Centres. Among the many names associated with St. Antony's mentioned in my recommended short reading list are the late Bill Deakin and Dick Storry, particularly for their brilliant work on the ace spy, Richard Sorge.

My last book of purely military history, *How Far from Austerlitz?*, was published nearly twenty years ago, but it left me with an enduring indebtedness to David Mynett, who accompanied me to the field of Austerlitz and sketched most of the illustrations. David was as talented a historian as he was an artist, and he helped me to see a great battlescape in terms of paint and perspective: particularly, in *Hubris*, the grim scenes before Moscow, 1941.

For specific items of research, I am indebted to Alistair Berven of Princeton University, for his work on Japan between the wars.

Among publishers I owe thanks to Alan Samson and the indefatigable George Weidenfeld for their initial enthusiasm for the *Hubris* project. I have been blessed with two of the brightest editors any author could ask for, Bea Hemming at Weidenfeld & Nicolson and Jonathan Jao at HarperCollins in New York. Then there is this mystical and magical figure, the copy editor. The title "copy editor" always seems to me to be an absurd diminution of the role they actually play in an author's life. Peter James, supreme in his calling, has that rare capacity to put himself inside the author's psyche: you disobey him at your peril. In the seven books that Peter has edited for me, which he calls "Horneana," I have tried hard not to.

If any error should nevertheless have eluded these editorial eagle eyes, it is, of course, the author where the buck stops.

For picture and map collation, my gratitude extends to Holly Harley.

I am beholden to my literary agents, Michael Sissons and Fiona Petherham of PFD in London and Peter Matson in New York, all of whom have provided constant support.

The toilsome task of marshalling my research notes and typing chapters has fallen to Liz Longley; I am particularly grateful for her ineffable good humor.

I owe warm appreciation in a different coin, notably, to two dedicated South African "carers," Zelda and Cathy, who brought me back from the dark abyss of a stroke—just in time to complete *Hubris*.

Finally, a tribute to my patient wife, Sheelin. Though she says she hates war, she has put up with it, sacrificing many an hour of her precious painting to read displeasingly martial manuscripts. Her support alone makes this book worthwhile—at least to me.

SELECT BIBLIOGRAPHY

Part I: Tsushima, 1905

Beasley, W. G. *Japanese Imperialism, 1894–1945* (Oxford, 1987).

———. *The Modern History of Japan* (London, 1963).

Buruma, Ian. *Inventing Japan, 1853–1964* (New York, 2003).

Falk, Edwin A. *Togo and the Rise of Japanese Sea Power* (New York, 1936).

Forczyk, Robert. *Russian Battleship vs. Japanese Battleship: Yellow Sea, 1904–1905* (Oxford, 2009).

Hough, Richard Alexander. *The Fleet That Had to Die* (New York, 1960).

Jukes, Geoffrey. *The Russo-Japanese War, 1904–1905* (Oxford, 2002).

Kennedy, Paul. *The Rise and Fall of the Great Powers: Economic Change and Military Conflict from 1500 to 2000* (New York, 1987).

Koenig, William. *Epic Sea Battles* (London, 1977; rev. ed., 2004).

Mahan, Alfred T. *The Influence of Sea Power Upon History, 1660–1783* (New York, 1890).

Morris, Edmund. *Theodore Rex* (New York, 2012).

Nish, Ian. *The Origins of the Russo-Japanese War* (London, 1985).

Pleshakov, Constantine. *The Tsar's Last Armada: The Epic Voyage to the Battle of Tsushima* (New York, 2002).

Seager, Robert. *Alfred Thayer Mahan: The Man and His Letters* (Annapolis, 1975).

Semenov, Vladimir Ivanovich. *The Battle of Tsu-Shima* (London, 1906).

———. *Rasplata, or the Price of Blood* (London, 1909).

Sondhaus, Lawrence. *Naval Warfare, 1815–1914* (New York, 2001).

Warner, Denis, and Peggy Warner. *The Tide at Sunrise: A History of the Russo-Japanese War, 1904–1905* (New York, 1975).

Westwood, J. N. *Russia Against Japan, 1904–1905* (Albany, 1986).

Part II: Nomonhan, 1939

Colvin, John. *Nomonhan* (London, 1999).

Coox, Alvin D. *Nomonhan: Japan Against Russia, 1939* (Stanford, 1985).

Deakin, F. W., and G. R. Storry. *The Case of Richard Sorge* (London, 1966).

Drea, Edward. *Nomonhan: Japanese-Soviet Tactical Combat, 1939* (Kansas, 1981).

Erickson, John. *The Soviet High Command: A Military-Political History, 1918–1941* (London, 2001).

Zhukov, Georgy. *G. Zhukov, Marshal of the Soviet Union: Reminiscences and Reflections* (Moscow, 1974).

Part III: Moscow, 1941

Billington, James H. *Fire in the Minds of Men* (New York, 1980).

Braithwaite, Rodric. *Moscow 1941: A City and Its People at War* (London, 2006).

Cohen, Eliot, and John Gooch. *Military Misfortunes: The Anatomy of Failure in War* (New York, 1990).

Deakin, F. W., and G. R. Storry. *The Case of Richard Sorge* (London, 1966).

Downing, David. *The Moscow Option: An Alternative Second World War* (London, 2001).

Erickson, John. *The Road to Stalingrad: Stalin's War with Germany*, Vol. I (London, 1983).

Erickson, John, and David Dilks. *Barbarossa: The Axis and the Allies* (Edinburgh, 1994).

Glantz, David. *Barbarossa: Hitler's Invasion of Russia, 1941* (Stroud, 2001).

———. *Colossus Reborn: The Red Army at War, 1941–1943* (Kansas, 2005).

———. *The Initial Period of War on the Eastern Front, 22 June–August 1941* (London, 1993).

———. *Stumbling Colossus: The Red Army on the Eve of World War* (Kansas, 1998).

Glantz, David, and Jonathan House. *When Titans Clashed: How the Red Army Stopped Hitler* (Edinburgh, 2000).

Gorodetsky, Gabriel. *Grand Delusion: Stalin and the German Invasion of Russia* (New Haven, 1999).

Grossman, Vasily. *Road to Victory: Twelve Tales of the Red Army* (London, 1945).

———. *A Writer at War: Vasily Grossman with the Red Army, 1941–1945*, eds. and trs. Antony Beevor and Luba Vinogradova (London, 2005).

Jukes, Geoffrey. *The Defence of Moscow* (London, 1969).

Krivosheev, G. F. *Soviet Casualties and Combat Losses in the Twentieth Century* (London, 1997).

Larionov, V. *World War II: Decisive Battles of the Soviet Army* (Moscow, 1984).

Lucas, James. *War on the Eastern Front: The German Soldier in Russia, 1941–1945* (London, 1998).

Overy, Richard. *Russia's War, 1941–1945* (London, 2010).

Pipes, Richard. *The Russian Revolution, 1899–1919* (London, 1990).

Pleshakov, Constantine. *Stalin's Folly: The Secret History of the German Invasion of Russia, June 1941* (Boston, 2005).

Sakharov, Andrei D. *Memoirs*, translated by R. Lourie (London, 1990).

Sebag Montefiore, Simon. *Stalin: The Court of the Red Tsar* (London, 2003).

Sella, Amnon. *The Value of Life in Soviet Warfare* (London, 1992).

Shukman, Harold. *Stalin's Generals* (London, 1993).

Stolfi, R. *Hitler's Panzers East: World War II Reinterpreted* (Oklahoma, 1992).

Volkogonov, Dmitri. *Triumph and Tragedy*, translated by Harold Shukman, 2 vols. (Moscow, 1989; London, 1991).

Werth, Alexander. *Moscow War Diary* (New York, 1942).

———. *Russia at War, 1941–1945* (London, 1964).

Zhukov, Georgy. *G. Zhukov, Marshal of the Soviet Union: Reminiscences and Reflections* (Moscow, 1974).

Part IV: Midway, 1942

Agawa, Hiroyuki. *The Reluctant Admiral* (Tokyo, 1979).

Cook, Theodore F., Jr. "Our Midway Disaster," in Robert Cowley, ed., *What If? The World's Foremost Military Historians Imagine What Might Have Been* (New York, 1999).

Dowler, John W. *War Without Mercy: Race and Power in the Pacific War* (New York, 1987).

Edgerton, Robert. *Warriors of the Rising Sun: A History of the Japanese Military* (Boulder, 1997).

Fuchida, Mitsuo, and Masatake Okumiya. *Midway: The Battle That Doomed Japan* (Annapolis, 1955).

Hastings, Max. *Nemesis: The Battle for Japan, 1944–45* (London, 2007).

Healy, Mark, and David Chandler, eds. Campaign Series, *Midway, 1942: Turning Point in the Pacific* (London, 1993).

Hughes, Wayne. *Fleet Tactics: Theory and Practice* (Annapolis, 1986).

Kahn, David. "Codebreaking in World Wars I and II: The Major Successes and Failures, Their Causes and Their Effects," *The Historical Journal* 23, no. 3 (1980), 617–39.

Keegan, John. *The Price of Admiralty* (London, 1988).

———. *The Second World War* (London, 1997).

Lord, Walter. *Incredible Victory* (New York, 1967).

Marder, Arthur. *Old Friends, New Enemies: The Royal Navy and the Imperial Japanese Navy, II: The Pacific War, 1942–1945* (Oxford, 1990).

Morison, Samuel Eliot. *History of United States Naval Operations in World War II,* Vol. IV: *Coral Sea, Midway and Submarine Actions, May 1942–August 1942* (Edison, 2001).

———. *The Two-Ocean War: A Short History of the United States Navy in the Second World War* (Boston, 1963).

Nish, Ian. *Anglo-Japanese Alienation, 1919–1952: Papers of the Anglo-Japanese Conference on the History of the Second World War* (Cambridge, 1982).

Overy, Richard. *Why the Allies Won* (New York, 1995).

Parshall, Jonathan, and Anthony Tully. *Shattered Sword: The Untold Story of the Battle of Midway* (Dulles, 2005).

Prange, Gordon W., Donald Goldstein, and Katherine V. Dillon. *Miracle at Midway* (New York, 1982).

Schlesinger, James R. *Midway in Retrospect: The Still Under-Appreciated Victory* (Washington, D.C., 2005).

Toland, John. *The Rising Sun: The Decline and Fall of the Japanese Empire, 1936–1945* (New York, 1961).

Willmott, H. P. *The Barrier and the Javelin: Japanese and Allied Pacific Strategies, February to June 1942* (Annapolis, 1982).

Part V: Korea and Dien Bien Phu, 1950–1954

Fall, Bernard B. *Hell in a Very Small Place: The Siege of Dien Bien Phu* (Philadelphia, 1966).

Grauwin, Paul-Henri. *Doctor at Dien-Bien-Phu* (London, 1955).

Greene, Graham. *The Quiet American* (New York, 1956).

Halberstam, David. *The Coldest Winter: America and the Korean War* (New York, 2007).

Hastings, Max. *The Korean War* (London, 1987).

Horne, Alistair. *The Price of Glory: Verdun, 1916* (London, 1962).
Karnow, Stanley. *Vietnam: A History* (2nd ed., New York, 1997).
Logevall, Fredrik. *Choosing War: The Lost Chance for Peace and the Escalation of War in Vietnam* (Berkeley, 1999).
Morgan, Ted. *Valley of Death: The Tragedy at Dien Bien Phu That Led America into the Vietnam War* (New York, 2010).
O'Ballance, Edgar. *The Indo-China War, 1945–1954* (London, 1964).
Porch, Douglas. *The French Foreign Legion: A Complete History of the Legendary Fighting Force* (New York, 1991).
Rees, David. *Korea: The Limited War* (New York, 1964).
Roy, Jules. *The Battle of Dienbienphu*, translated by Robert Baldick (New York, 1965).
Sheehan, Neil. *A Bright Shining Lie: John Paul Vann and America in Vietnam* (New York, 1988).
Stueck, William W. *Rethinking the Korean War: A New Diplomatic and Strategic History* (Princeton, 2002).
Toland, John. *In Mortal Combat: Korea, 1950–1953* (New York, 1991).

Epilogue

Ashrawi, Hanan. *This Side of Peace: A Personal Account* (London, 1995).
Bregman, Ahron. *Israel's Wars: A History Since 1947* (2nd ed., London, 2002).
Carey, Roane, ed. *The New Intifada: Resisting Israel's Apartheid* (London, 2001).
Donovan, Robert J. *Israel's Fight for Survival* (New York, 1967).
Eban, Abba. *Abba Eban: An Autobiography* (London, 1977).
Heikal, Mohamed. *The Road to Ramadan* (London, 1975).
Herzog, Chaim. *The Arab-Israeli Wars: War and Peace in the Middle East* (London, 2010).
———. *War of Atonement* (Boston, 1974).
Horne, Alistair. *Kissinger:1973, The Crucial Year* (London, 2009).

———. *Macmillan, 1894–1956* (London, 1988).

Kissinger, Henry. *White House Years* (Boston, 1979).

Meir, Golda. *My Life* (London, 1975).

O'Ballance, Edgar. *The Arab-Israeli War, 1948* (London, 1956).

———. *The Third Arab-Israeli War* (London, 1972).

Oren, Michael. *Six Days of War: June 1967 and the Making of the Modern Middle East* (Oxford, 2002).

Ovendale, Ritchie. *The Origins of the Arab-Israeli Wars* (4th ed., Oxford, 2004).

Safran, Nadar. *From War to War: The Arab-Israeli Confrontation, 1948–1967* (New York, 1969).

Shlaim, Avi. *The Iron Wall: Israel and the Arab World* (London, 2000).

Weizman, Ezer. *On Eagles' Wings* (Tel Aviv, 1975).

LIST OF ILLUSTRATIONS

INDEX

ABOUT THE AUTHOR

SIR ALISTAIR HORNE is the author of over twenty books on history and politics. They include *A Savage War of Peace: Algeria, 1954–1962* (winner of the Wolfson History Prize); *The Price of Glory: Verdun, 1916* (winner of the Hawthornden Prize); *How Far from Austerlitz? Napoleon, 1805–1815*; and *Seven Ages of Paris*. In 1969 he founded the Alistair Horne Fellowship to help young historians at St. Antony's College, Oxford. He was awarded the French Légion d'honneur in 1993 and received a knighthood in 2003 for his work on French history. Horne and his artist wife, Sheelin, live in Oxfordshire.